PLANE AND GEODETIC SURVEYING

SECOND EDITION

PLANE AND GEODETIC SURVEYING

SECOND EDITION

Aylmer Johnson

CRC Press
Taylor & Francis Group
Boca Raton London New York

CRC Press is an imprint of the
Taylor & Francis Group, an **informa** business

A SPON PRESS BOOK

CRC Press
Taylor & Francis Group
6000 Broken Sound Parkway NW, Suite 300
Boca Raton, FL 33487-2742

First issued in paperback 2019

ISBN-13: 978-1-4665-8955-1 (hbk)
ISBN-13: 978-0-367-86824-6 (pbk)

Library of Congress Cataloging-in-Publication Data

Johnson, Aylmer, 1951-
    Plane and geodetic surveying / author, Aylmer Johnson. -- Second edition.
      pages cm
    Includes bibliographical references and index.
    ISBN 978-1-4665-8955-1 (hardcover : alk. paper)
    1. Surveying. I. Title.

TA545.J64 2014
526.9--dc23   2013038224

Visit the Taylor & Francis Web site at
http://www.taylorandfrancis.com

and the CRC Press Web site at
http://www.crcpress.com

To Zara and Chrissie—in the hope that, one day,
they too will discover the fun of this subject.

Contents

Preface to the Second Edition

In the 10 years since the first edition of this book was published, there have been a number of advances in surveying equipment: total stations have almost entirely replaced theodolites, GPS has been augmented by other satellite navigational systems, and the shaping of new roadways is now done by remotely-controlled earth-moving machinery using satellite receivers. More than this, though, it seems that I have learned more about the subject from 10 further years of teaching able and inquisitive students, who enjoy asking their lecturers difficult questions. The result is hopefully a book which is once again technically up-to-date, and which explains the subject with slightly more rigour and clarity than before.

Aylmer Johnson
Cambridge
May 2013

Preface to the First Edition

More than almost any other engineering discipline, surveying is a practical, hands-on skill. It is impossible to become an expert surveyor, or even a competent one, without using real surveying instruments, and processing real data. On the other hand, it is undoubtedly possible to become a very useful surveyor without ever reading anything more theoretical than the instrument manufacturers' operating instructions.

What, then, is the purpose of this book?

A second characteristic of surveying is that it involves much higher orders of accuracy than most other engineering disciplines. Points must often be set out to an accuracy of 5 millimetres with respect to other points, which may be more than 1 kilometre away. Achieving this level of accuracy requires not only high-quality instruments but also a meticulous approach to gathering and processing the necessary data. Errors and mistakes which are minute by normal engineering standards can lead to results which are catastrophic in the context of surveying.

Yet in the real world, errors will always exist, and approximations and assumptions must always be made. The accepted techniques of surveying have been developed to eliminate those errors which are avoidable and to minimise the effects of those which are not. Likewise, the formulae used by surveyors incorporate many assumptions and approximations, and save time when the errors which they introduce are negligible by comparison with the errors already inherent in the observations.

No two jobs in surveying are exactly the same. A competent professional surveyor therefore needs to know the scope and limitations of each surveying instrument, technique, and formula: partly to avoid using unnecessarily elaborate methods for a simple job but mainly to avoid using simplifying assumptions which are invalidated by the size or required accuracy of the project. This knowledge can only be developed by understanding how the accepted techniques have evolved, and how the formulae work—and this understanding is becoming increasingly hard to acquire with the advent of electronic 'black box' surveying instruments and software applications, which perform elaborate calculations whose details are hidden from the user.

It is this understanding which this book sets out to provide. The methods for using each generic class of surveying instrument have been described in a way which is intended to show why they have evolved; and the calculations are similarly explained, such that the inherent assumptions can be clearly identified. Wherever necessary, practical guidance is also given on the range of distances for which a particular formula or technique is both necessary and valid.

The material in this book is based on the surveying courses taught in the Engineering Department at Cambridge University, and I am grateful to the many colleagues who have both enhanced my own understanding of the subject and contributed to past editions of the 'Survey Notes', from which this book evolved. The philosophy of engineering education at Cambridge has always been that an understanding of a subject's fundamental principles is the key to keep abreast with the changes which technology inevitably brings; and indeed to initiating appropriate changes, when technology makes this possible. I hope that this book has succeeded in applying that philosophy to surveying, in a way which will be of value to those who read it.

Aylmer Johnson
Cambridge
August 2003

Acknowledgements

I am indebted to the many colleagues and students who have helped to shape the various drafts of this text, and the accompanying computer program. In particular I am grateful to two of my former colleagues, both sadly now deceased: John Matthewman, who sparked my interest in surveying and taught me much of what I now know about it; and Wylie Gregory, who initially wrote the core algorithms within the adjustment program. My brother Bill also deserves special mention for his valuable suggestions while proofreading this edition.

About the Author

Aylmer Johnson is a senior lecturer in the Cambridge University Engineering Department, and a Fellow of Clare College. His main research interests are in the area of knowledge management in engineering design, but he has also taught surveying throughout his academic career, and has led the Surveying Group at Cambridge University for the last 10 years.

Chapter 1

Introduction

1.1 AIM AND SCOPE

Engineering works such as buildings, bridges, roads, pipelines and tunnels require very precise dimensional control during their construction. Buildings must be vertical, long tunnels must end at the correct place, and foundations must often be constructed in advance to accommodate prefabricated structural sections. To achieve this, surveyors are required to determine the relative positions of fixed points to high accuracy, and also to establish physical markers at (or very close to) predetermined locations. These tasks are achieved using networks of so-called control points. This book aims to give the civil engineering surveyor all the necessary theoretical knowledge to set up, manage and use such networks for the construction and monitoring of large or small engineering works.

The exact way in which control networks are established and managed depends on a number of factors:

1. *The size of the construction project and the accuracy required.* The accuracy of each technique described in this book is explained, together with the limitations of the various assumptions used in subsequent calculations. In particular, guidance is given as to when a project is sufficiently large that the curvature of the earth must be taken into account.
2. *The available equipment.* As far as possible, the descriptions of surveying equipment in this book are generic and are not based on the products of any one particular manufacturer. Both satellite and 'conventional' surveying instruments are covered, since both are appropriate under different circumstances.
3. *The country in which the work is being carried out.* This book explores some topics with particular reference to the mapping system used in Great Britain; but a clear indication is also given of how the same issues are addressed in other countries, with different mapping systems and survey authorities.

The tools of the engineering surveyor have changed significantly in recent years. Most notably, navigational satellites now provide the simplest and most accurate way of finding the position of any point on the surface of the earth, or (more importantly) the relative positions of two or more points. This first became commonplace in the mid-1980s, with the advent of the global positioning system, or GPS; other systems are now also available, and the method is collectively known as the global navigational satellite system, or GNSS. However, GNSS has some inherent limitations, which are explained in detail in Chapter 7. As a result, other more conventional surveying techniques must still be understood, and used when appropriate. In addition, the traditional surveying tools such as levels and total stations are now predominantly electronic, and can usually record (and sometimes even observe) readings quite automatically. The descriptions in this book are largely independent of these advances, because they do not change the basic way in which the instrument works, but simply mean that readings can be taken more simply and quickly. The particular techniques for using manual instruments have also been included, for those occasions and locations where electronic instruments are not available or appropriate.

1.2 CLASSIFICATION OF SURVEYS

Surveys are conducted for many different purposes, which will determine the types of instruments which are used, the measurements which are taken, and the subsequent processing of those measurements to produce the required results. It is useful to know the names of the principal types of surveys, and the nature of the work which is involved.

Engineering surveys are usually classified in the following ways:

1.2.1 Classification by Purpose

1. *Geodetic*—to determine the shape of the earth, or to provide an accurate framework for a big survey, whose size means that the curvature of the earth must be taken into account;
2. *Topographic*—to produce ordinary medium-scale maps for publication and general use. Topographic surveys record all the features of the landscape which can be shown on the scale of the map. Topographic maps are usually produced by means of aerial or satellite photogrammetry.
3. *Cadastral*—to establish and record the boundaries of property or territory. Cadastral surveys are concerned only with those features of the landscape which are relevant to such boundaries.

4. *Engineering*—to choose locations for, and then set out markers for, engineering construction works. Engineering surveys are concerned only with the features relevant to the task in hand, and usually have two phases. The first phase typically involves collecting an amount of topographic information, to allow the project to be planned in detail. When the planning is complete, the second phase consists of setting out the necessary markers for earthmoving and construction to begin.

Engineering surveyors are likely to come in contact with all these types of surveys, but are less likely to have to conduct a large-scale topographic survey. For this reason, the techniques of topographic surveying are not covered in this book; for further information on this class of surveying, see Mikhail, Bethel and McGlone (2001) and Wolf (2000).

1.2.2 Classification by Scale

The scale (i.e., size) of a survey will affect the instruments and techniques used, as well as the type of projection used to display or record the results. In a topographic survey, it also determines the amount and type of topographic detail which is recorded.

1.2.3 Classification by the Type of Measurements Taken

1. *Triangulation*—finding the size and shape of a network of triangles by measuring their angles and (since about 1980) the lengths of their sides. Used in conventional surveying when each station can see three or more other stations.
2. *Traverse*—proceeding from one point to another by 'dead reckoning', using measured distances and angles to calculate bearings. Used when the construction work is long and narrow, such as a motorway or tunnel.
3. *Resectioning*—establishing the precise position of a single station by measuring distances and angles to a number of other nearby stations whose positions are already known.
4. *DGNSS* (differential GNSS)—measuring the relative three-dimensional (3-D) positions of two stations by simultaneously recording satellite observations at each one, and comparing the results.

Often, a particular survey involves a hybrid of all four of these techniques.

1.2.4 Classification by the Equipment Used

1. *Tape*—for direct linear measurement. Cheap and robust. Still occasionally used for small detailed surveys, but now largely supplanted by electromagnetic distance measurement devices.
2. *Compass*—to observe bearings. Used mainly in preliminary reconnaissance.
3. *Theodolite*—a telescopic sight pivoted horizontally and vertically, with two graduated protractors (called 'circles') for measuring angles. See Figure 1.1.
4. *Electromagnetic distance measurement (EDM)* devices—typically used for measurements of lengths from, say, 5 m to 5 km, though some instruments have ranges up to about 25 km.
5. *Total station*—essentially a theodolite with a built-in EDM. Total stations usually have facilities for recording and processing measurements electronically, and have largely replaced conventional theodolites.
6. *LiDAR*—acronym which stands for light detection and ranging, and refers to a class of surveying instruments which can measure distances and angles to solid surfaces to a reasonable accuracy at a very high rate. A LiDAR receiver generates a three-dimensional 'cloud' of points and is an efficient way of capturing a three-dimensional shape, e.g., the surface of some land, the façade of a building, or the details of a chemical plant. These devices are also known as laser scanners.
7. *GNSS*—using navigational satellites to fix positions on the earth. This technique has almost completely replaced terrestrial triangulation for

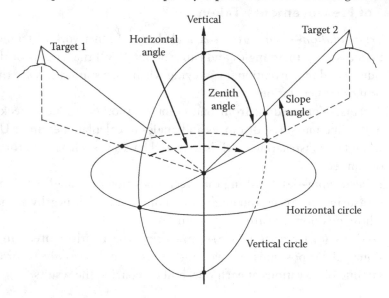

Figure 1.1 Angles measured by a theodolite or total station.

large-scale control survey, and can also be useful on building sites, provided it is not set up close to buildings or trees. The term 'satellite surveying' is also used for this activity, but is ambiguous—see 9, below.

8. *Aerial camera* (photogrammetry)—mainly used in topographic surveys, but also for recording the shapes (and subsequent deformations) of buildings: see Atkinson (2001).
9. *Satellite camera*—essentially, a long-range aerial camera. Satellites can be used for gathering topographic data, and also for many other remote sensing purposes related to geographic information systems (GIS), such as monitoring crop yields or atmospheric pollution.

1.3 THE STRUCTURE OF THIS BOOK

There are several possible ways to set out the further theoretical knowledge that a competent engineering surveyor will need. The approach adopted in this book is to describe the most important *principles* of surveying in Chapter 2, and the main activities of surveyors in Chapter 3. This is followed by four further chapters explaining how particular measurements are made: angles in Chapter 4, distances in Chapter 5, heights in Chapter 6, and satellite position fixing in Chapter 7. The use of satellite data in particular requires an understanding of geodesy, including the shape of the earth, and the co-ordinate systems used to describe the positions of points on or near to its surface; these concepts are discussed in Chapters 8 and 9. Chapters 10 and 11 cover the main calculations a surveyor will need to process raw observations into useful results; and Chapter 12 describes some additional conventional methods for finding height differences which are particularly useful when dealing with tall structures, or when establishing local transforms for raw GNSS data.

The appendices contain useful numerical data, observation and calculation sheets, and worked examples of some of the calculations described in the book. A glossary is also included, explaining terms which might be unfamiliar to the reader.

Finally, a least-squares adjustment program called LSQ is available on the book's webpage (www.crcpress.com/product/isbn/9781466589551). LSQ will adjust a mixture of conventional and differential GNSS observations to compute the most likely positions of stations whose positions are unknown, and the likely accuracy to which they have been found.

Chapter 2

General Principles of Surveying

Surveying has two notable characteristics: the work is done to a much higher level of accuracy than most other engineering work; and it is possible for quite serious errors to remain undetected until it is too late to correct them. For this reason, there are some inherent principles which should be observed in all surveying, regardless of the type of survey or the equipment used. This chapter describes those principles.

2.1 ERRORS

All the results of surveying are based on measurements, and all measurements are subject to errors. Because surveying involves high degrees of accuracy (most surveying measurements are accurate to within 10 parts per million and some are within 1 or 2 parts per million), it is relatively easy to make significant errors, and relatively hard to detect them. The understanding and management of errors is therefore possibly the single most important skill that a professional surveyor must possess. Many of the techniques of surveying are directed towards cancelling or eliminating errors, and towards ensuring that no serious error remains undetected in the final result. Even so, the presence of unnoticed 'systematic' errors in a survey can lead to false, yet seemingly consistent, results. A recent international tunnelling project drifted several metres from its intended path because temperature gradients near the tunnel wall caused lines of sight to bend consistently in one direction, and this was not detected until an independent method was used to check the work.

High accuracy in surveying is expensive because it involves costly, high-quality equipment and more elaborate procedures for taking measurements. On the other hand, cheaper equipment may not be adequate to achieve the required accuracy, particularly if (for instance) a long distance has to be split into several steps, requiring more measurements and resulting in an accumulation of errors. Surveys are therefore often conducted using high-quality equipment to establish a few 'major control' stations around the area to a higher accuracy than is required overall, and then filling in the intervening detail by cheaper methods adequate for the shorter distances. This is usually the most economic way of distributing the 'error budget' to achieve a satisfactory final result at minimum cost.

2.1.1 Types of Errors

Surveying errors fall into three categories:

1. *Gross errors.* Gross errors are due to mistakes or carelessness, such as misreading by a metre or a degree. A proper routine of checks should detect them. A surprisingly common source of error is the manual transcription of readings from one place to another.

2. *Systematic errors.* Systematic errors are cumulative and due to some persistent cause—generally in an instrument, but sometimes in a habit of the observer. They can be reduced by better technique but not by averaging many readings, as they are not governed by the laws of probability. Thus all distances measured with an inaccurate tape or electromagnetic distance measurement (EDM) device will, from that cause, have the same percentage or absolute error, whatever their lengths and however many times they are measured; the only remedy is to calibrate the device more carefully. This is the most serious sort of error, and the technique of survey is mainly directed against it: the greater the accuracy required, the more elaborate and expensive the instruments and the technique.

   A special type of systematic error is a *periodic* error, which varies cyclically within the instrument. Examples include errors in the positions of the angle markers on a horizontal circle, inaccuracies in a vernier scale, or nonlinearities in the phase resolver of an EDM or global navigational satellite system (GNSS) receiver. This type of error can sometimes be eliminated by special observation techniques: for instance, measuring a horizontal angle several times, but using a different part of the horizontal circle on each occasion.

3. *Random errors.* Random errors are due to a number of small causes beyond the control of the observer. Their magnitude depends on the quality of the instrument used and on the skill of the observer, but they cannot be corrected. Thus no one can place a mark, or make an intersection, or read a scale with absolute accuracy or consistency. Even after allowing for systematic personal bias (covered in 'Systematic errors' above), there will remain errors which are a matter of chance and are subject to the laws of probability. In general, positive and negative errors are equally probable; small errors are more frequent than large ones, and very large random errors do not occur at all.

In statistical terms, random errors cause readings to deviate from the correct value in the manner of a normal distribution—similar, for instance, to the scatter of heights to be found in a sample of adults. The scale of the scattering can therefore be defined by quoting the *standard deviation* (σ) of the distribution; two-thirds of all readings will lie within one standard deviation of the correct value (above or below), and 95% within two standard deviations. Alternatively, the standard deviation for the random element in an observation can be estimated by taking the measurement several times, and seeing what range of values covers the middle two-thirds of the readings; the size of this range is an estimate of ($2 \times \sigma$).

Two other measures of quality are also used to define the accuracy of readings affected by random errors. The *probable error,* expressed as $\pm p$, is such that 50% of a large number of readings differ from the correct value by less than p; for normally distributed errors, p is 0.675 times the standard deviation of the readings. A more useful measure of accuracy is the 95% *confidence value* which, as explained above, is almost exactly two standard deviations.

Assuming that the observation errors from an instrument have a normal distribution (i.e., that they contain no gross or systematic errors), it can be shown that the standard deviation associated with the arithmetic mean of a set of n repeated observations is $1/\sqrt{n}$ times the standard deviation of a single observation. Thus if a single angle measurement can be read to one second of arc, the mean of four readings should have a precision of 0.5 seconds. Taking the same measurement several times can therefore be a valid way of increasing the overall accuracy of a survey.

2.1.2 Precision and Accuracy

In understanding the nature of measurement errors, it is important to appreciate the distinction between *precision* and *accuracy*. It is possible, for instance, to measure a distance to fairly high precision (0.5 mm or better) using a simple tape measure—but if the marks have not been printed in the right places on the tape, the reading will not be accurate. Even when the greatest precautions are taken in making a reading (e.g., measuring the distance again, using a different part of the tape), systematic errors (e.g., the whole tape has become longer because of thermal expansion) may still dominate the results. Caution must therefore be used when estimating the standard deviation of a set of observations from the apparent 'scatter' of the results, as described above. A set of consistent readings indicates a consistent instrument and a good observer, but not necessarily an accurate result.

It is good practice to avoid recording observations to a higher precision than is warranted by their accuracy. However, this is not always done, and surveyors should be aware that observational data is sometimes considerably less accurate than it appears to be.

2.2 REDUNDANCY

Given two points whose positions are known, the position of a third point in plain view can be found by (for instance) measuring the horizontal distances between it and the two known points. However, the accuracy of the calculated position can only be inferred from the quoted accuracy of the distance measurement device; and a gross error in one of the distance measurements (or an error in the quoted position of one of the known points) will still give a seemingly plausible solution for the new point's position.

To overcome both of these problems, a fundamental principle of surveying is to take redundant readings: that is, to take more measurements than are strictly necessary to fix the unknown quantities. Any large inconsistency in the readings will then indicate a gross error in the measurements or the data, while any small inconsistencies will give an unbiased indication of the likely accuracy to which the point has been fixed.

When several new points are to be fixed simultaneously, it can become quite difficult to ensure by simple inspection that enough suitable readings have been taken or planned to ensure redundancy throughout the network. This soon becomes apparent, though, when the readings are adjusted by computer (see Section 2.4 below). For this reason, many adjustment programs include a planning mode, which enables a proposed scheme of observations to be validated for redundancy before it is carried out. A surveyor is strongly advised to carry out such a check, if there is any doubt about the redundancy of a proposed scheme of observations.

2.3 STIFFNESS

In addition to being redundant, a network (and its associated observations) should also be 'stiff'—in other words, the relative positions of control points and the scheme of observations should be arranged such that any significant movement of one of the points would cause a correspondingly significant change in at least one of the observations. This ensures that the positions of unknown points are established to the highest possible accuracy, using the instruments which are available.

There is an exact analogy (as with redundancy) between a 'stiff' network and a stiff structure. The pin-jointed structure shown in Figure 2.1(a) is stiff, because any given deflection of point C requires that member AC or BC (or both) must lengthen or shorten by a similar amount. In Figure 2.1(b), by

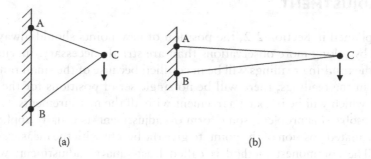

Figure 2.1 **Stiff and non-stiff structural frameworks.**

contrast, the structure is much less stiff since C can make quite large vertical movements with relatively small changes in the lengths of the two members.

The corresponding situation in surveying is shown in Figure 2.2, where points A and B are known points, and C is unknown; and the distances AC and BC have been measured. As with the structure, Figure 2.2(a) shows a stiff network, in which any significant movement of point C would involve equally significant changes to one or both of the measured distances; whereas in Figure 2.2(b), C could move significantly in the north/south direction without greatly affecting either of the distances.

If angle measurements are used as well, this corresponds to adding gusset plates to the structure, which increases its stiffness by removing the freedom in the pin joints.

As with redundancy, it can be quite difficult to determine by inspection whether a proposed scheme of observations will result in a stiff network. Again, though, an adjustment program with a 'planning' facility will provide a good prediction of how accurately the unknown points will be fixed, if the likely accuracy of the planned observations is known.

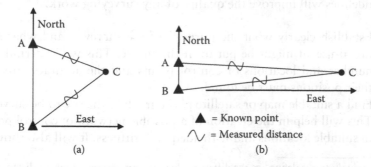

Figure 2.2 **Stiff and non-stiff survey networks.**

2.4 ADJUSTMENT

As explained in Section 2.2, the position of new points should always be found by taking more observations than are strictly necessary. Inevitably, then, the resulting readings will be in conflict; because of the small random errors in the readings, there will be no single set of positions for the new points which will be in exact agreement with all the measurements.

To resolve this problem, some form of 'adjustment' is usually applied to the calculated position of the point, to give the best fit with the measurement data. The commonest method is called least-squares adjustment, which chooses positions for the new points such that the sum of the squares of the residual errors[*] is minimised. This gives the most likely positions for the new points, assuming that the observation errors are normally distributed.

A good understanding of what adjustment can, and cannot, achieve is important for a surveyor. Essentially, it is a statistical process which gives the most likely position for each new point, assuming that the observation errors are random and normally distributed. If this is not the case, the results may be misleading or inaccurate. In particular, least-squares adjustment will give a false impression of accuracy if there are systematic errors present in the data, e.g., if all distance measurements are made using a device which is poorly calibrated. It will also generate misleading results if the user is tempted to reject any seemingly 'bad' observations purely on the grounds that they do not appear to agree well with the others.

Adjustment is described in greater detail in Chapter 11.

2.5 PLANNING AND RECORD KEEPING

A successful survey requires an appropriate set of measurements to be taken and recorded without unnecessary deployment of human resources or equipment. This can only be achieved by means of planning. The following guidelines will improve the quality of any surveying work.

1. Establish clearly what the purpose of the survey is and what additional uses it might be put to in the future. This will determine the number and locations of control points and the accuracy to which their positions must be found.
2. Find a suitable map or satellite photograph of the site to be surveyed. This will help in the creation of a possible network of control points, in suitable locations and with adequate stiffness. It will also show the

---

[*] The residual error is defined as the difference between an observed angle or distance, and the calculated value based on the assumed position(s) of the new point(s).

approximate scale of the work and will help in detecting gross errors in angle and distance measurements.

3. Visit the site if at all possible. Check whether control stations can be sited at the places indicated by Step 2, and make a note of what will be needed to build them. If conventional instruments are to be used, check whether the necessary lines of sight exist between the station locations, using ranging rods if necessary. If GNSS is to be used, check that the relevant stations have a clear view of the sky. Make notes of any features on the site (cliffs, ditches, etc.) which might make it difficult to move from one station to another.

A few simple instruments may also help at this stage. A compass can be used to estimate horizontal angles, and a clinometer will measure approximate vertical angles. A hand-held GNSS receiver will give the approximate co-ordinates of points and estimates of the distances between them. If this is not possible, a hand-held laser measure can be used, or the distances can be paced.

4. Plan a set of observations which will establish the control network to the required accuracy at minimum cost. This is generally best done by working 'from the whole to the part': accumulated errors are minimised by first forming an accurate framework covering the whole area, and then adding further control stations to whatever accuracy is necessary. Accurate measurements require expensive equipment and longer observation times, so this type of consistent approach will give the most economical result.

The planning function in an adjustment program is very useful here. The eventual quality of a network can be reliably predicted by entering *approximate* observations (such as the compass angles above), together with estimates of the accuracy to which the final measurement will be made.* Different observations can then be included in the scheme, to see which combination will give an adequate accuracy for minimum investment. Make sure, though, that there are enough observations so that one or more could be rejected without unacceptable loss of accuracy or redundancy. The time spent travelling to and from a site is usually much greater than that needed to take a few 'spare' measurements while an instrument is set up.

5. Plan the fieldwork in detail to make sure that all the necessary measurements are taken with the minimum deployment of people and equipment. Each member of the team should know who will take which measurements, at which locations, and with what instruments.

---

* The approximate observations establish the geometry of a network to sufficient accuracy for its eventual stiffness to be determined. This, combined with the accuracy of the final observations, determines the accuracy to which the points in the network will eventually be fixed.

6. If possible, arrange that all fieldwork has redundancy, and that the computations are carried out such that no gross error (Section 2.1.1) will pass undetected. If some of the error checks can be carried out in the field while the equipment is still set up on station, then the cost of correcting any error will be greatly reduced.

7. Before leaving base, make sure that all batteries are fully charged and that any necessary co-ordinate data, transformations, etc., have been downloaded into those instruments that need them. Make sure that everyone is familiar with the instruments they will be using: get unfamiliar instruments out, read the instruction manuals, and practise their use.

8. Ensure that each group of surveyors keeps a diary of what is done, including a summary of the weather, on each day. If an error is discovered later, a good diary can be invaluable in pinpointing the source of the problem, and thus showing which measurements may need to be repeated.

9. Make sure that observation records are complete, and will not degrade with time—the data generated during a surveying job may need to be consulted many years after it was initially made. Observations recorded on paper should be checked for legibility and completeness, and stored in a dry condition; electronic data should be stored on a long-term medium, such as a CD-ROM. For important jobs, copies of the data should be made and stored in a different location from the originals—the cost of this is minuscule compared to the cost of taking the measurements again. Finally, a brief summary of the data will greatly assist any subsequent attempt to re-inspect some part of it.

Chapter 3

Principal Surveying Activities

Many surveying books start by explaining how the various instruments are used, and then describe the reasons for their use. This order has been reversed here, for the benefit of those who like to understand the ultimate purpose of a technique before studying it in detail. Some of the concepts mentioned in this chapter may therefore not be fully clear to readers who are new to this subject; such readers are encouraged to refer to later chapters, as necessary.

3.1 ESTABLISHING CONTROL NETWORKS

Before any survey can yield useful results, it is necessary to establish a set of fixed stations whose positions relative to one another are known—usually to a higher accuracy than will be needed in the final result. A set of such stations is known as a *control network*.

If the scope of an engineering project is relatively small (up to 5 km square, say) and does not have to be tied in with work elsewhere, then it is usually easiest to set up a local Cartesian co-ordinate system for the work, and to use conventional surveying instruments rather than a global navigational satellite system (GNSS). Typically, the first control station* is established at or near the southwest corner of the site, and defined to be the 'site origin', having the co-ordinates $(0,0,0)$†. A second station is then set up at the northwest corner of the site with its x co-ordinate defined to be 0. The horizontal line between the two stations defines the y-axis, or 'site north', and the z-axis is defined to be vertically upwards. An orthogonal Cartesian co-ordinate system is thus fully specified, such that any point on the site has a unique (x,y,z) or (easting, northing, height) co-ordinate.

---

* See Appendix B for a full discussion of control stations.
† Often a set of positive co-ordinates is chosen for the site origin, e.g., (100,100,100) so that no point on the site has negative co-ordinates.

At least one further control station will also be needed on the site; each additional station is set up by first choosing a suitable location, then physically establishing the station, and finally taking measurements to find its co-ordinates. The two-dimensional (2-D) (x,y) position of each station is found by measuring horizontal angles and/or distances to or from other stations (see Chapters 4 and 5). If needed, the height (z) co-ordinates are usually found separately by levelling, as described in Chapter 6.

If just one further control station is to be added to the initial two points, there would be three unknowns in the 2-D co-ordinate system: namely, the y co-ordinate of site north, and the (x,y) co-ordinates of the third control station. Finding these unknowns with redundancy thus requires at least *four* measurements, of which at least *two* must be horizontal distance measurements (if one distance and three angles were measured, there would be no check that the distance had been measured correctly). A typical scheme of measurements for fixing a third station is shown in Figure 3.1: here, a horizontal angle has been measured at Station 1, and the instrument has then been moved to Station 3, where a second angle and two distances have been measured.

If the final network is to consist of more than three control stations, then a minimum of $(2n - 2)$ readings is required to achieve redundancy in two dimensions, where n is the total number of stations (i.e., including the site origin and site north). In addition, redundancy considerations require that:

1. There should again be at least two distance measurements;
2. Site north (which has one unknown) should be involved in at least *two* measurements; and
3. Each subsequent point (which will have two unknowns) should be involved in at least *three* measurements.

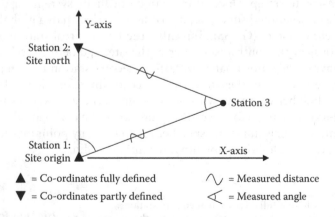

Figure 3.1 Survey network with three control stations.

These requirements at least ensure that no gross error will pass unnoticed—but it is generally advisable to take several additional measurements over and above this minimum, so that any problematic reading can be eliminated from the set altogether, without loss of redundancy.

The total number of control stations required in the network, and their relative positions, will depend on the size of the site and the purposes for which they are needed. If the intention is to set out further points on the ground whose own relative positions must be guaranteed to be accurate (e.g., the foundation points for a prefabricated bridge), then there should ideally be three or more control stations near to each point, arranged so that the positions of the new points will be sufficiently accurate and totally error-proof when they are set out. If the purpose includes the production of some type of map, then each relevant feature of the landscape must be visible from (and not too far from) one of the control stations. Further control stations may also be needed, simply to ensure that the relative positions of the 'useful' stations are known to a sufficient degree of confidence and accuracy—and also to ensure that the site co-ordinate system will not be lost if one or both of the original stations is destroyed or displaced.

Many variants exist for establishing a local Cartesian system of the type described above. There is no need for 'site north' to be the same as true north, though it reduces the likelihood of mistakes if they are more or less in the same direction. Likewise, the station which defines the 'site origin' does not have to be the lowest point, or at the south-west corner of the site—but if it is not, then the chances of gross errors are again reduced if its co-ordinates are not (0,0,0) but are defined such that every point on or around the site has positive co-ordinates.

In larger surveys, it is often necessary or more convenient to use an existing regional co-ordinate system, or grid. This is typically done by using nearby existing control stations with known co-ordinates, to find the grid co-ordinates of the main control stations around the site. Such grid systems are usually orthogonal, but they often involve a scale factor; this means that one metre in the grid system does not exactly correspond to one metre of horizontal distance on the ground. On a very large survey, this scale factor will alter between one place and another; and this causes discrepancies between angles observed in the field and those measured between the corresponding straight lines drawn on the grid projection (see Chapter 9 for further details). There are also complications in using a national datum for height measurements, which are explained in Chapter 8.

3.1.1 Satellite Methods

Since about 1980, the most straightforward way to find the relative positions of stations has been to use differential GPS (or more recently, DGNSS). Navigational satellite receivers are simultaneously placed on two different stations, and their relative positions are known to within about 5 mm after approximately half an hour's 'observation'. If national grid co-ordinates are required, one or more national GNSS control points (whose grid co-ordinates are now typically published over the Internet) are also included in the scheme of observations. A particular advantage of using differential GNSS is that the stations do not need to have a line of sight between them. The use of GNSS is explained in Chapter 7.

GNSS has not, however, completely supplanted the more traditional ways of establishing control networks. It cannot be used if the control stations need to be near tall buildings, beneath trees, or in tunnels—and the parameters required to transform the data into a ground-based co-ordinate system can only be checked by making conventional measurements between some of the control points as well. The equipment is also relatively expensive and is potentially subject to undetectable systematic errors if used on its own; so again, it is reassuring to have an independent method of checking the results it produces. The remainder of this section therefore describes the more traditional ways of establishing control networks.

3.1.2 Triangulation

Until about 1970, nearly all control networks were created by a process called *triangulation*. Two stations were established on the ground, and the distance between them (called the 'base line') was carefully measured* (usually in both directions). The relative position of a third station could then be fixed (with partial redundancy) by measuring all three angles in the triangle between it and the other two stations. No further distance needed to be measured, which was an advantage in the days when distances could only be measured by tape. More stations could subsequently be added (with redundancy) to the network by measuring angles between them and three or more of the stations already in the network.

For a given instrument accuracy and time budget, the best overall control network for a particular area using triangulation is obtained by distributing control stations as evenly over the area as possible, with well-conditioned† triangles, and all stations visible‡ from at least three others. Since computing power was also expensive prior to 1970, the total number of stations was kept as low as possible at this stage; if further control was subsequently

---

* Accuracy is crucial here, since any error in this measurement will cause an undetectable 'scale error' to propagate through the whole network.

† No angle in the triangle should be less than about 20° or greater than about 160°.

‡ Two stations cannot be regarded as being visible from each other if the line of sight between them passes close to ('grazes') some piece of intervening ground. This will have the effect of bending the light path between them, greatly reducing the accuracy of angle and distance measurements.

required in part of the area, more stations could be established nearby, and fixed with reference to the existing stations.*

The advent of electromagnetic distance measurement (EDM) in about 1970 made distance measurement much easier and cheaper, and meant that many of the sides of the triangles in such networks could now be measured as well. This has the effect of making any network much 'stiffer', and eliminates the possibility of an undetected scale error. However, the use of distances to fix the grid positions of stations even in two dimensions requires knowledge of the altitudes of the endpoints, as will be shown in Chapter 10—so the use of distance measurements is often kept to a minimum in conventional surveys, even now.

3.1.3 Traversing

When the area to be controlled is long and thin (e.g., a tunnel or a motorway), or when each station can only see two others, a system of interlocking triangles is impracticable, and a so-called traverse is used instead. In its simplest form this consists of setting up a total station† over a station whose co-ordinates are known, observing another station (with known co-ordinates) as a reference object, then observing the horizontal angle and distance to a station whose position is unknown, but which can now be calculated from the information available. The instrument is now set up over the new station, and the process is repeated for each 'unknown' station in turn, finishing up on a final 'known' station. The agreement between the calculated co-ordinates and the known co-ordinates for this final station is a measure of the accuracy of the traverse; and there are two pieces of redundancy in the set of observations, which can be used to give improved estimates of the positions of all the unknown stations in the traverse.

In practice, most control networks are now established by some hybrid of triangulation, traverse, and GNSS. The key features of a good network are that it should be both stiff and fully redundant, as described in Chapter 2, and that, if possible, it should be free of any systematic error.

---

* This was the logic which dictated the construction of the first-order network over the United Kingdom by the Ordnance Survey between 1936 and 1951. There are approximately 480 first-order stations covering Great Britain, most of which are at the tops of hills or mountains; in flat parts of the country, church towers and water towers are used instead. A much larger network of second-order stations was subsequently established, using the first-order stations as fixed points; and these in turn provided fixed points for establishing third-order stations.

† The glossary gives an explanation of this term and of others which may be unfamiliar.

3.2 MAPPING

Most civil engineering projects require the input of surveyors at two different stages—firstly to find out what currently exists on a site, and later to establish suitable markers for the new constructions that have been planned. The making of maps is often necessary at the first of these stages.

This book does not attempt to cover mapping to the depth required by cartographers; however, engineering surveyors will often need to make precise maps of small areas, so a few useful guidelines are given here.

1. Decide what scale of map is required before starting. On a 1:1000 map, a line of width 0.2 mm will represent 0.2 metres on the ground, so there is no value in recording the positions of points to better than this accuracy.
2. For the same reason, there is no value in recording details of shapes which are too small to show up on the map. If a length of fence or hedge does not deviate from a straight line by more than one line's width when plotted on the map, then only the positions of the end-points need be recorded. (The eventual line on the map may of course be drawn as a wavy line, to show that the detail on the ground is not exactly straight.)
3. The purpose(s) of the map will determine what details need to be recorded and how this should be done. If the map is to be used to plan the positions of new control points, then the line which records, say, the edge of a ditch should mark the closest point to the ditch where it would be sensible to establish a control point, rather than the water's edge. Likewise, the diameter of a tree's trunk might be of relevance for determining lines of sight on the ground, while the size of its canopy may be of relevance for GNSS observations.
4. If the map is to be made back at base, it is useful to draw a freehand sketch of the intended map before starting to take measurements. This will indicate the amount of detail which can usefully be recorded, and the sketch can be marked up with numbers which relate to the points whose positions are actually measured.
5. If points are to be recorded by means of a total station, it will first be necessary to establish control stations such that every point of interest can be observed from one control point or another. Also, each control station used for the mapping must have a line of sight to some other known point, which can be used as a reference bearing for the other measurements. If a control station is only needed for mapping purposes, then its position probably doesn't need to be known to better than decimetre accuracy—but it may be wise to establish it to higher accuracy than this, in case it is subsequently needed for some other purpose.

6. To make a map, the instrument is set up on one control station and sighted on another, which acts as a reference object. Some instruments can then use the co-ordinates of the instrument position and reference object to 'orient' themselves, and calculate co-ordinates for all subsequent sightings—otherwise, the raw data is recorded and processed back in the office. A staff-holder then takes a detail pole (with a reflector) to each point of interest, and the bearing, vertical angle and slope distance are recorded by the instrument. So-called robotic total stations have stepping motors on their horizontal and vertical axes, and can follow the reflector automatically: the surveyor with the detail pole can then tell the instrument when to take the readings, by means of a radio link. This means that the job of collecting detail can be done by a single surveyor—but there have been cases of robotic total stations being stolen, when the surveyor is too far away to prevent it!

7. It is important to keep the detail pole vertical during measurements, so that its tip (which is resting on the ground) is directly below the reflector. If the detail pole is adjustable, its length is set to be the same as the height of the instrument (i.e., the trunnion axis) above its control station. This simplifies the mapping of heights, as the measured height difference from the instrument to the reflector can simply be added to the (known) height of the instrument's control station to find the height of the feature that the detail pole is resting on.

8. Small areas are now often mapped using real-time kinematic (RTK) differential GNSS. The operator walks from one point of interest to another, and records their positions using a GNSS receiver mounted on a detail pole. An electronic map can be produced simultaneously by joining the points with curves or straight lines and adding symbols or descriptive text, using a hand-held computer.

9. Sometimes, the techniques described in (6) and (8) above are used simultaneously. A robotic total station is set up at a suitable (but arbitrary) position, and detail is collected by means of a detail pole equipped with a reflector *and* a GNSS receiver. The total station reads the angle and distance to the reflector, and also records the computed position of the GNSS receiver via a radio link. When the detail pole has been moved to a few different locations, the total station is able to compute its own position and orientation from this data; then, if GNSS reception is not possible at some of the places that the detail pole is taken to, its position can be calculated using the angle and distance measured by the total station.

As well as making land maps, surveyors sometimes need to 'map' complex shapes such as the façade of a building or the steelwork of a bridge. This is now typically done using a LiDAR, or laser scanner; this can be thought of as a high-speed robotic reflectorless total station* which can systematically scan and measure in any direction except (usually) downwards, and produces a three-dimensional (3-D) 'point cloud' of observations, at rates of up to 1 million points per second. These points are fed into software which drapes a surface over the cloud and produces a computer model of the object. Most LiDAR scanners also incorporate a conventional digital camera, so that the surface of the model can be automatically rendered with the correct colours.

If (for instance) a whole building is to be mapped in this way, it becomes necessary to move the LiDAR sensor to different places and collect a separate point cloud from each location. If there is some overlap of detail between the various point clouds, the software is able to 'stitch' the clouds together and produce a complete, solid model. This principle can be carried further: a vehicle equipped with a LiDAR scanner and a GNSS receiver can drive down a road and map all the visible details on either side, or an aircraft can create a detailed 3-D terrain model of the land it flies over.

3.3 SETTING OUT

The most common ultimate purpose of an engineering survey is to 'set out' points at predetermined positions, either on the ground or on partly-built structures—e.g., to mark where soil should be moved to, where foundation points should be built (perhaps for a prefabricated construction), or where the columns should be located on a partly-completed building for its next storey.

For much setting-out work, the accuracy does not need to be greater than a centimetre or so, which allows quicker, less precise techniques to be used. Redundancy is often not required either—if a large number of points are being set out (e.g., to mark out the centre line of a curved road or railway track), it is generally obvious when one point has been put in the wrong place. This makes the process quite simple: if differential GNSS is being used, the required co-ordinates are typed in and the system guides the surveyor to the correct place. If a robotic total station is used, the process is very similar: the surveyor holds the detail pole and faces the instrument, which then transmits instructions of the form 'x metres closer and y metres to your left' to a display on the pole, to steer the surveyor to the correct location. When a manual total station is used, the process is essentially the reverse of a mapping exercise: the required co-ordinates are used to compute a distance from one control point, and a horizontal angle with respect to a second control point. To fix the point, a total station is set up on the first control point, sited onto the second control point, and turned through

---

* Some robotic total stations can be programmed to function in this way too; they are generally more accurate than a LIDAR scanner, but much slower.

the calculated angle. A detail pole is then moved along the telescope's line of sight, until the correct distance is measured.

Having located a point by one of these methods, a short wooden peg might be driven into the ground, with a nail in its top surface to mark the exact point. If two such pegs are set out, a straight line can be defined on the ground by means of a taut string between the two nails. If an exact height is needed, a taller post is driven into the ground and a level is used to mark a line of collimation (which will have a known height)* on the post, or on a taller ranging rod beside it. A horizontal board, called a sight rail, is then nailed to the post with its upper surface at the desired height.

The advent of GNSS has brought about some major changes in the techniques of setting out, which are still in the process of rapid evolution. Most modern bulldozers, for instance, are equipped with at least two GNSS receivers (e.g., one on the cab roof and one on the top of the front blade) plus a digital radio link and servo controls; this allows a central computer to monitor the position and orientation of the bulldozer's blade continuously, and thus create a soil surface at exactly the required height without the need for any sight rails or other physical markers on the site. This is both quicker and safer than the more conventional methods—surveyors do not need to go into sites where large vehicles may be operating to set out physical markers.

There is thus a wide and still-growing range of setting-out procedures for different engineering purposes. It is beyond the scope of this book to cover them all—and many of them are very well and fully described both in Schofield and Breach (2007) and in Uren and Price (2006). Instead, the remainder of this section describes how to set out single points to the highest possible accuracy; and Section 3.4 describes how to assess that accuracy, once the location has been fixed.

3.3.1 Setting Out in the Horizontal Plane

If the point is in a suitable place for GNSS observations† and the transformation between the GNSS and local co-ordinate system has been established, then real-time kinematic GNSS can be used as described above to guide the operator to the desired point, usually to an accuracy of a centimetre or so. Prolonged observation at that provisional point will then determine its position to a higher accuracy, and an appropriate small movement can be made, if necessary, to improve the location of the point.

If GNSS is not available or appropriate, new points can most easily be set out in the horizontal plane by calculating their bearings and distances from an existing control station whose co-ordinates are known. A total station is set up on the existing station and sighted onto a second 'known' control

---

* See Chapter 6 to learn how this is done.

† See Chapter 7 Section 7.4 for an explanation of why some places are unsuitable for this.

station, which acts as a reference object. The angle to be turned through, and the distance to be measured*, are calculated by simple geometry; see Appendix F for a worked example of these calculations. The instrument is turned through the appropriate angle, and a target is moved along the telescope's line of sight until the measured distance is correct, which places it at (or at least close to) the desired point. Many total stations can perform these calculations automatically, given the co-ordinates of the two control stations and the new point, and will make an audible noise when the telescope is pointing in the right direction and again when the target is in the correct place.

For full accuracy, this process needs to be done twice—once with the instrument in the face 1 configuration, and again in face 2 (see Chapter 4 for an explanation of what 'face 1' means, and why this is necessary). This results in two points being set out which should be fairly close to each other—the 'best guess' for the correct location of the point is then midway between them.

When isolated points are being set out for major construction work, it is essential to have some degree of redundancy, so that any error will be detected before the work starts. The simplest form of redundancy is to set the point out again, using a different method or (if only total stations are available) different control points as reference points. If the two resulting points are reasonably close to each other, the point halfway between them can be used, as described above; if the two points are *not* close to each other, this indicates a gross error which needs to be investigated further.

An alternative method, which generally avoids the use of distances, involves setting up total stations (or theodolites) over three nearby control points (not necessarily simultaneously) and sighting them along the relevant bearing lines towards the set-out point, as described above. A small 'setting-out table' is fixed in approximately the correct position, and the lines of sight (face 1 and face 2) from the three instruments are plotted on its surface, to give a figure similar to that shown in Figure 3.2.

Generally, the lines will not cross at a point, due to errors in the positions of the three control points, and in setting up the three lines of sight. However, they should very nearly do so—the sides of the triangle should not be greater than a centimetre or two. Assuming that the error is likely to be distributed equally between all three lines, the most likely position of the required point is at the centre of the inscribed circle, as shown in the figure, and the radius of the circle gives an indication of the likely accuracy to which the position of the point has been established.

The following points should be noted, when using this method:

---

* Remember that, for high accuracy, distances may need to be corrected for the local scale factor of the grid and for the heights of the two stations; see Chapters 9 and 10.

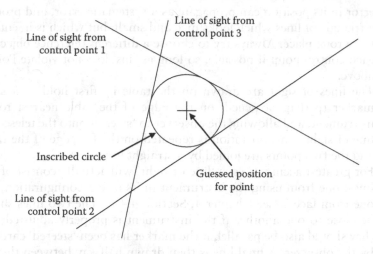

Figure 3.2 Establishing the best position for a set-out point.

1. Three lines of sight from existing control stations are necessary to guard against the possibility of a gross error, e.g., in calculating one of the bearings. Two nonparallel lines of sight will always cross at a point; if the third one also passes nearby, it is reasonably unlikely that any gross error has occurred.

2. The three control stations should be positioned such that the triangle formed by the three lines of sight is a reasonably well-conditioned one, as shown in Figure 3.2. If two lines of sight cross each other at a very acute angle (less than about 30°), then the accuracy of the final point will be degraded, as it would be strongly affected by a small movement of either of the two lines. Note, however, that this does *not* require the control stations to be spaced at near-120° intervals round the set-out point, as each station can be at either end of the line of sight.

3. The three control stations should be reasonably close to the set-out point. As well as making it easier to walk between the instrument and setting-out table and instrument, this reduces the effect of any inaccuracy in the instrument's sight line; if this inaccuracy is (say) 5 seconds, this would cause an error of about 1 mm over a distance of 50 metres, but 25 mm over a distance of 1 kilometre.

4. By the same reasoning, the reference object should be as far away from the control point as possible, and at least as far away as the point to be set out. Otherwise, any inaccuracies in the relative positions of the reference object and control point will cause a larger inaccuracy in the setting out.

5. It is good practice not to use the same reference object from all three control points. If the same reference object is used throughout, any

error in its position can propagate as a systematic error, and produce a triangle of lines which looks nice and small, but which is in entirely the wrong place. Always try to choose a different reference object for each control point if possible, so long as this does not violate Point 4 above.

6. The lines of sight are drawn on the table by first holding a small marker (perhaps a pencil) on the edge of the table nearest to the instrument, and allowing the observer to 'steer' it onto the telescope's line of sight. This operation is repeated on the far edge of the table, and the two points are joined by a straight line.

7. For greatest accuracy, each 'line of sight' will actually consist of two lines: one from using the instrument in its face 1 configuration, and one from face 2 (see Chapter 4, Section 4.4.1). The two lines should lie close to one another if the instrument is properly adjusted; and they should also be parallel, if the marker has been 'steered' carefully by the observer. A final line is then drawn halfway between the two observed lines and is taken as the 'line of sight' from that instrument.

8. It is not necessary to have three instruments for this task. The table can be positioned, and the first two lines drawn, using two instruments set up on two of the control stations. When this has been done, one of the instruments is moved to the third control station, and the final line is drawn on the table. If only one total station is available, the table can be positioned using the line of sight and distance from the first control point, and the first line can then be drawn before the instrument is moved. If just one theodolite is available, the first 'line of sight' (from just one face) can be approximately recorded using two ranging rods, with a piece of string tied between them. The theodolite is then moved to the second control station, and the table is positioned where the second line of sight crosses the taut string. Lines are now drawn on the table from the second station, and the theodolite is then moved back to the first station and on to the third station for the other lines.

9. If three well-conditioned sets of lines cannot be sighted onto the table, then two sets are drawn, and a suitable target is set up above the intersection. The horizontal distance to the intersection is then measured from one control station and compared with the calculated distance. A third straight line can then be drawn on the table perpendicular to the corresponding line of sight and the appropriate distance from the intersection, to give a locus of all points on the table which are the 'correct' distance from that control station. An inscribed circle can then be drawn to touch the three lines, as described above.

10. When the most likely position of the point has been established on the setting-out table, a tripod can be set up over it and an optical or laser plummet sighted down onto the mark. The table can then be carefully removed and a more permanent marker established at the point where

the plummet sights onto the ground. (The drawing from the table should be kept as part of the record of the work.)

11. The size of the triangle gives an indication of the accuracy to which the point has been set out, but only in relation to the points which were used in the setting out: remember to add in the inaccuracies of those points if the absolute accuracy of the set-out point is required. For an independent check of the work, and a better indication of its accuracy, the point should be resectioned, as described in the next section.

Whichever method of setting out is used, it is important to remember that a systematic error can give an apparently acceptable result which is, in fact, in the wrong place. An error in computing the positions of the local control points could have affected them all equally, and an error in specifying a GNSS-to-grid transformation could give incorrect, yet quite consistent, GNSS results. For important setting-out points, it is good practice to use as independent a method as possible to check the results. A mixture of GNSS and total stations might be used or, if four points have been set out to form the rectangular base of a building, the lengths of the sides and diagonals could be measured as a simple check. Additionally, of course, the point(s) can be resectioned—preferably, using some control points which were not used in the initial setting-out.

3.3.2 Setting Out Heights

This is a relatively easy process, compared to setting out in the horizontal plane. For greatest accuracy, a temporary benchmark (TBM) is established at the place where the height is to be set out, using the techniques described in Chapter 6. When the height of the TBM is known, a staff or tape can be used to measure up (or down) to the required height, where a horizontal wooden sight rail might be fixed. If the accuracy of the rail's height is critical, it should be checked independently once it has been established, ideally by levelling to a staff resting on the top of the rail.

On building sites, height control is often achieved by means of a precisely horizontal laser beam which rotates about a vertical axis. This provides a constant height datum across the entire site, by painting a horizontal line on anything placed in its path. A similar technique can be used to ensure that steelwork, for instance, is erected vertically: here, the axis of rotation is set to be horizontal, so the rotating laser beam defines an exactly vertical plane. Two such beams can be set up approximately at right angles to each other, to ensure that (for instance) a concrete column cast on site is exactly vertical.

Satellite surveying (i.e., GNSS) is not generally used for setting out heights accurately, because it tends to be slower and less accurate than the conventional methods. Usually, a set-out height needs to be accurate only with respect to some nearby datum, and this is much more quickly and easily achieved using a level.

3.4 RESECTIONING

When a point has been set out and the final monument of its position has been established, it is often prudent to carry out further checks to ensure that the mark is in the correct place. The process of finding the exact horizontal location of a single point with respect to other known stations is called resectioning.

If the area is suitable for satellite surveying, and if the point was initially set out by conventional means, then the obvious way of carrying out this task is by means of differential GNSS (see Chapter 7), as this provides a completely independent check of the point's location.

However, if the point was initially established using real-time kinematic GNSS, it is good practice to use a conventional method to check it, as this will guard against the possibility of a systematic error in the GNSS observations or the data processing. This should involve the measurement of angles and (especially) distances with respect to nearby known stations. The process may be as simple as using a total station to check that the two foundation points for a bridge are the correct distance apart, or it may require all the set-out points to be independently resectioned into the control network.

To confirm the position of a point in two dimensions* with some degree of redundancy, three or more measurements will be required. This could be any combination of horizontal distances and horizontal angles—but note that measuring three horizontal angles from a point involves observing to four different control stations.

Normally, resectioning is done by taking all the necessary measurements from the point whose position is to be determined. If (for instance) three control stations were used to set a point out initially, then two horizontal angles can be measured by observing those three stations; one or more of the distances to the stations can be measured as well, to provide redundancy.

Care is sometimes needed to ensure that the measurements which are taken are well-conditioned—i.e., that they will fix the position of the point to the best possible accuracy. With distance measurements, this is usually fairly obvious: if a point was resectioned by measuring distances to stations which were all nearly due east or west of it, there would be considerable uncertainty about its position in the north–south direction. In the case of angle measurements, the effect is more subtle, as shown in Figure 3.3. If

---

* Height would normally be treated separately; see Chapter 6.

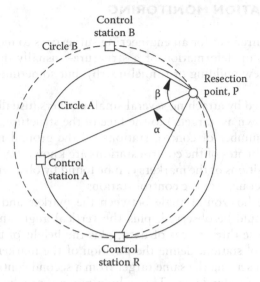

Figure 3.3 Poorly conditioned resection point.

the angle RPA is measured, a piece of information is obtained which will help fix the point P. Specifically, P is now known to lie somewhere on the circle labelled A, which passes through P, R and A. However, P could lie anywhere on this circle, and the same angle (α) would still be observed. Likewise, if the angle RPB were observed, P would be known to lie somewhere on circle B, which passes through P, R and B. If this was the same circle as circle A (i.e., if P, R, A and B all lay on a single circle) then nothing further would be known about the position of P, despite the additional measurement. If (as shown) control station B does not quite lie on circle A, then the position of P will be fully determined by the second measurement—but only to a very low degree of accuracy compared to the accuracy of the angle measurements. For instance, measurements taken from any point near P on circle A would yield an identical value for α, and an almost identical value for β.

Normally, the measurements taken to fix the position of P would be fed into a least-squares adjustment program (see Chapter 11) to find its most likely location. If the measurements were poorly conditioned, this would become obvious from the results, which would show a large 'error ellipse' for the position of P. The site would then have to be visited again to take further measurements, which is clearly undesirable. Surveyors should therefore be aware of the conditions under which measurements are likely to be ill-conditioned, even at the earlier stage of deciding where to position control points which might be used for subsequent resectioning.

3.5 DEFORMATION MONITORING

A common requirement for an engineering surveyor is to monitor the possible movement or deformation of a structure—usually during the construction of a new building or tunnel nearby, but sometimes over a much longer period.

This is achieved by attaching several small markers (usually small reflective stickers known as 'targets') to the face of the structure, and establishing a suitable number of control stations on the ground in the vicinity. Once the co-ordinates of the control stations are known (including relative heights), the positions of the markers can be found by observing them using a total station set up over the control stations.

Typically, the horizontal angle between the marker and a known reference point would be observed, plus the vertical angle and distance* to the marker. These three pieces of data, plus the height of the instrument above the control station, define the position of the marker (but with no redundancy). Observing the same target from a second control station will provide adequate redundancy. This whole process can be (and often is) completely automated, by having permanently-mounted robotic total stations which are programmed to search out and measure a group of markers at regular intervals.†

If the distance to the marker cannot be measured, then observing the marker from two stations will still precisely define (in three dimensions) where it must be—namely, where the two lines of sight intersect. In fact, there is even a degree of redundancy, as the two lines will generally not quite intersect each other—so the most likely position for the marker is at the midpoint of the shortest link between the two lines, and the length of the link is an indication of the accuracy. Generally, though, it would be wise to obtain further redundancy (and further accuracy) by observing from a third station, under these circumstances.

Deformation monitoring is typically the most precise type of surveying, and often involves measuring the positions of points to the nearest 0.1 mm. All procedures therefore need to be carried out with special care, particularly the setting up of instruments exactly above station marks, and measuring their exact heights above those marks. If a nonrotating optical or laser plummet is used to set the instrument up above its mark, then its accuracy should be checked at regular intervals (see Chapter 4, Section 4.5.2).

The following guidelines are also useful:

---

* Even if the target is reflective, this would generally require the use of a reflectorless total station.
† These can often be seen in London's underground stations.

1. The control stations should, if possible, be no more than about 50 metres from the markers, to prevent unpredictable errors caused by the tendency of a light path to bend in the atmosphere (see Chapter 10, Section 10.2).

2. Particular attention should be paid to the 'error ellipses' generated when the initial positions of the markers are found through least-squares adjustment. Their sizes will indicate the smallest subsequent movement which can be detected with confidence.

3. A surveyor must consider (and possibly take advice on) the possibility that some of the control stations may be affected by the same movements which affect the structure. If this is unavoidable (e.g., to comply with the 50 metre rule, above), then additional control stations must be established further away, and the 'vulnerable' stations resurveyed before subsequent measurements are made on the structure.

4. The surveyor must also be clear what types of movement need to be measured when setting up the control network. It may or may not be necessary to detect movements which affect all the nearby terrain as well as the structure itself. If it is, then further, more distant control points must also be established. Ultimately, GNSS provides the best detection of movements of large areas of terrain, even including continental drift.

5. Deformation monitoring using adhesive markers can be used in conjunction with photogrammetry to monitor changes in shape of the structure (or rock face, etc.) at points away from the markers. If the markers are visible in the photographs, the position of any other distinguishable feature can be found using this technique; see Atkinson (2001). However, this technique has now been largely supplanted by laser scanning.

6. For long-term work, it is advisable to have many more markers than necessary on the structure, as some are likely to be lost with the passage of time; and, of course, there should be enough control stations to ensure that the reference co-ordinate system is not lost if one or more of them is destroyed.

Chapter 4

Angle Measurement

A number of the techniques discussed in Chapter 3 require the measurement of angles, to a greater or lesser degree of accuracy. Horizontal and vertical angles can be measured approximately using a compass and clinometer, respectively; for more accurate work, a total station or theodolite will be needed. This chapter explains in detail how angles are measured using these instruments.

4.1 THE SURVEYOR'S COMPASS

The standard surveyor's compass is a hand-held device which shows the bearing of a line relative to magnetic north. A graduated circular card incorporating a bar magnet rests on a low-friction pivot; prisms or mirrors and sights are arranged so that the graduations on the card may be read whilst making a sighting on the distant point. Damping is incorporated, and there is usually a locking device for the card whilst the instrument is not in use. Bearings may be read to 0.5° (or 1 part in 120, when the angle is converted to radians).

The angle subtended by two stations at a third one can thus be estimated to within a degree by taking the magnetic bearings of the two stations from the third, and subtracting one reading from the other.

Note, however, that the individual bearings are shown with respect to magnetic north, rather than true north (where all the meridians meet), or grid north. The difference between these two bearings can be up to 25° on the main land masses of the world (and of course up to 180° near the poles) and is discussed further in Chapter 9.

4.2 THE CLINOMETER

In its simplest form, a clinometer consists of an optical sighting system with a pendulum attached to it. The pendulum has a protractor attached to it, so that the inclination of the sight line can be measured. A distant object is observed through the sights, and a prism enables the protractor to be read at the same time, giving the vertical angle* of the line between the observer's eye and the distant object to approximately 20 seconds. When using a clinometer, be sure to note whether a zenith angle is being read (i.e., the angle made with the vertical) or a slope angle (i.e., the angle made with the horizontal).

A variation on the clinometer is the sextant, which is typically used at sea to measure the vertical angle of the sun, or of another star or planet. Here, the horizon is used as a reference direction instead of a pendulum, and the optics of the device allow the difference in vertical angles to be observed. This is more useful on a boat at sea where (a) the horizon always defines a near-horizontal reference direction, and (b) a pendulum would be likely to oscillate.

4.3 THE TOTAL STATION

4.3.1 The Instrument

Horizontal and vertical angles (see Figure 1.1) are measured accurately using a total station (or a theodolite, which is a similar instrument but which lacks the ability to measure distances). Instruments are classified by the way in which angles are read—total stations are nearly always digital, and theodolites are generally optical. They are also classified by the standard deviation of error which can be expected from a reading; thus, a half-second instrument will give angle readings that have a 95% likelihood of being accurate to 1 second (about 1/200,000 of a radian, i.e., 5 parts per million).†

Angles are normally measured in degrees (360 to one complete rotation), minutes (60 to one degree) and seconds (60 to one minute). However some instruments measure in gons (400 to one complete rotation) and decimal fractions (0.001 gon = 1 milligon, 0.0001 gon = 1 centesimal second). Some instruments also measure in radians (2π radians to one complete rotation) and decimal fractions (e.g., milliradians). Radians used to be avoided in optical instruments, since there is not a rational number of radians in a full circle; this is less of a problem with digital instruments. The following discussion assumes an instrument which works in degrees, minutes and seconds.

In construction, the instrument is a telescopic sighting device (described below) mounted on a horizontal axis (called the trunnion axis) whose

---

* The meaning of this term is explained more fully in Section 4.3
† Note that, on optical instruments, the classification sometimes refers to the precision to which the instrument can be read and *not* to the accuracy of the angle which has been obtained as a result.

bearings are in turn mounted on a vertical axis. Thus the telescope can be pointed freely in any direction. In particular, the telescope may be rotated through 180° about the trunnion axis without altering any other setting; this is known as transiting. A protractor, or 'circle', is mounted in a plane perpendicular to the trunnion axis, and to one side of the telescope, for measuring vertical angles; when the telescope, viewed from the eyepiece end, has this circle on the user's left, it is said to be in the 'circle-left' (CL) position; if the telescope is then transited, it will be in the 'circle-right' (CR) position. On most digital instruments, however, it is impossible to see where the vertical circle is; such instruments therefore have a way of displaying a I or II to the observer, to show whether they are in 'face 1' (F1) or 'face 2' (F2) configuration; F1 always corresponds to CL, and F2 to CR. Because most instruments are now digital, F1 and F2 will be used to describe an instrument's configuration in the rest of this chapter.

For vertical angles, it is conventional to measure the *zenith angle*; this angle is zero when the telescope is pointing vertically upwards, 90° when it is horizontal in the F1 position, and 270° when horizontal and in F2. However, most digital instruments can also be configured to show a *slope angle*, which is zero when the telescope is horizontal, 90° when it is pointing vertically upwards, and –90° when it is pointing vertically downwards (on either face).

Each motion, horizontal and vertical, may have a clamping screw; when this is tight (*never* more than finger-tight), a tangent screw provides a limited range of fine adjustment. Some instruments use a friction system in place of a clamp—this makes the instrument or telescope slightly harder to rotate, but avoids the need to tighten a clamp before the tangent screw will work.

The instrument as a whole may be levelled using a spirit level so that its "vertical axis" is in fact exactly vertical; the horizontal axis is constructed so as to be accurately perpendicular to the vertical axis, though provision is made for adjustment if this becomes necessary. Usually, the instrument is mounted on a tripod, which allows it to be placed directly above a fixed marker on the ground: such centring is normally carried out using either a plumb-bob, or an optical or laser plummet. Often the instrument is mounted on a tribrach, which in turn is mounted on the tripod.

The telescope of a total station or theodolite is an aiming device with four essential parts:

1. The object-glass, the optical centre* of which is effectively the foresight of the telescope's sighting system.
2. The reticle, usually a glass diaphragm, carrying an engraved cross, with a horizontal and a vertical line, which is set to be on the optical

---

* A lens acts in the same way as a pinhole, except that it collects more light and has a specific focal length; the optical centre of the lens can be thought of as the position of the equivalent pinhole.

axis of the telescope, and which forms the backsight of the sighting system.

3. An internal focussing lens which is used to focus, in the plane of the reticle, the image of the target formed by the object-glass.

4. The eyepiece, which magnifies the reticle and the image of the target, and allows them to be viewed comfortably.

The exact form of the reticle lines (commonly called hairs since on early instruments that is what they were) engraved on the diaphragm varies, but most instruments have a single full-width horizontal line and two shorter horizontal lines, called the stadia lines, one above and one below the full-width line.* For all vertical readings, the central horizontal line is used. Vertically, some instruments have a single full-depth central line, and others have a single line on one side of the horizontal line and a pair of lines on the other (see Figure 4.1).

A very common error when sighting the telescope vertically is to use one or the other stadia line instead of the central full-width line. Always make sure that you can see *all* the lines engraved on the reticle before taking a reading.

For angle readings, an engraved disk or 'circle' is mounted on each axis. In digital instruments, these circles are engraved with bar codes which are scanned by an optical reader and converted to digital angle readings. In an optical instrument, the circles are pieces of glass engraved with degrees and minutes, which are read directly by the observer; the reading system incorporates a vernier (usually optical) to enable the full accuracy of the measurement to be obtained.

As shown in Figure 1.1, horizontal angles are measured with respect to a reference object, so always involve subtracting one horizontal angle reading from another. This process is straightforward on a digital instrument—the displayed horizontal angle is set to zero by pressing a button when the telescope is pointing at the reference object, and the instrument then does all the necessary subtractions internally. On an optical instrument, the horizontal circle may itself be rotated about the vertical axis, so that the angle

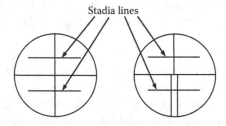

Stadia lines

Figure 4.1 Typical reticle designs.

---

* The purpose of these lines is explained in Chapter 5, Section 5.3.

reading to the reference object can be set to a convenient value; the circle must then, of course, remain undisturbed relative to the axis throughout each subsequent reading.

The vertical circle also rotates about its (horizontal) axis, so that its zero can be set to the vertical. This may either be done manually, with the aid of a special spirit level (called the alidade bubble) attached to the circle, or (now more commonly) automatically, using a damped pendulum.

A special type of theodolite, called a gyrotheodolite, incorporates a gyro-scope which is able to align itself with the spin axis of the earth. This means that (except near the north or south pole) the instrument is able to read an azimuth* angle as its horizontal angle, without the need for a physical reference object. Gyrotheodolites are specialist instruments which need additional training to use, but they are valuable for ensuring that a long traverse (e.g., down a tunnel) has not deviated from its measured direction as a result of refraction effects. See Schofield and Breach (2007) for further details.

4.3.2 Handling

Total stations and theodolites, whether optical or digital, are delicate and expensive instruments. They should be handled with great care. Do not jar or knock them, or use the slightest force on them. Do not touch or rub the lenses; if these get wet, they can be blotted gently with a tissue.

In the field, the instrument is normally put in its container for every move, and not carried on its tripod. If (for short moves in steady conditions) it is ever carried on its tripod, see that it is secure on the tripod, not too far up on its foot-screws, and that any clamps are just tight. When carrying an instrument in this way, keep it vertical; it is not designed to resist stresses at any other angle.

When being transported in a vehicle, these instruments are vulnerable to high-frequency vibration, which can cause large accelerations even at small amplitude. It is good practice to transport them in a foam surround, or with the container held on the lap of a passenger.

Take special care if the instrument has to be set up, even temporarily, on a hard surface. The legs of a tripod are very liable to slip, and should be placed in a 'spider' or chained together for safety. Never leave an instrument unattended—children and animals are very inquisitive. The instrument should always be protected from rain and, for precise work, from the sun.

---

* The angle with respect to the meridian on which the instrument is located. All meridians meet at the north and south poles which, in the case of a gyrotheodolite, are defined to be where the spin axis of the earth passes through its surface.

4.3.3 Setting Up, Centring and Levelling

There is no one best way of centring and levelling a tripod. The method described below is simple and works well for centring over a station mark on reasonably level ground. The ability to set up tripods quickly and accurately over a station mark is necessary when working with targets and global navigational satellite system (GNSS) receivers, as well as total stations.

1. Position the tripod so that its top is roughly horizontal and above the station mark, using a plumb-bob if desired. The top should be approximately level—this can be achieved without affecting its position over the mark by moving the tripod feet *tangentially*, rather than radially. On sloping ground, it is better to have one leg directly uphill and, if the legs are adjustable, shorter than the others.

2. Attach the instrument to the tripod; if it has a detachable tribrach with its own plummet, this may conveniently be used on its own for the initial adjustments. The instrument or tribrach should be set at the centre of the tripod head, with the foot-screws about halfway along their travel. The optical or laser plummet will now be roughly vertical, and should point within one or two centimetres of the ground mark: if it is more than about 5 cm from the mark, recheck the orientation of the tripod head and make sure it is approximately horizontal.

3. If necessary, focus the eyepiece of the optical plummet so that the reticle and station mark both appear sharp (some optical plummets have separate focussing rings for these two functions). If your tripod has legs whose length can be adjusted, tread the feet of the tripod firmly into the ground at this stage.

4. Adjust the levelling screws of the instrument or tribrach until the station mark is on the crosshairs of the optical plummet—or, if using a laser plummet, until the mark is illuminated by the laser.

5. Level the cup bubble on the instrument or tribrach by adjusting the lengths of one tripod leg followed by another, without moving the feet. As shown in Figure 4.2, this allows the tribrach to be levelled while hardly altering the point on the ground observed by the plummet. If the tripod has fixed-length legs, you can achieve the same effect by moving one foot at a time *radially*, as they have not yet been trodden in. You should find that the plummet is still very nearly pointing at the station mark; if it is not, repeat Steps 4 and 5.

6. Fully tighten the leg clamps if the tripod has adjustable legs, or tread in the feet if it has fixed-length legs. Make final levelling adjustments to the cup bubble on the instrument or tribrach using the levelling screws, and then use the tripod centring adjustment to bring the optical or laser plummet onto the station mark. If the centring adjustment has insufficient travel, you will have to return to Step 4.

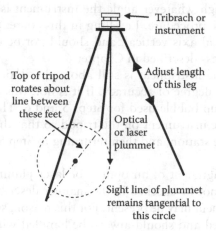

Tribrach or
instrument

Top of tripod
rotates about
line between
these feet

Adjust length
of this leg

Optical
or laser
plummet

Sight line of plummet
remains tangential to
this circle

Figure 4.2 Centring a tribrach over a ground mark.

7. If the plummet is built into the instrument rather than the tribrach, rotate the instrument through 360° and watch the position of the sighting point or laser spot on the ground as you do so. If it moves in a circle, this indicates that its line of sight is not aligned with the vertical axis of the instrument; the effect of this misalignment can be eliminated by moving the instrument on the tripod so that the station lies at the centre of the circle.

8. For targets, GNSS antennae and self-levelling instruments*, the process is now complete. If manual levelling is necessary, rotate the instrument so that the horizontal plate bubble (physical or electronic) lies parallel with an imaginary line between two of the foot-screws. Centre the plate bubble by turning these foot-screws equally in opposite directions—you will find that the bubble follows your left thumb. Swing the instrument† through 90° and centre it again by turning the third foot-screw only. Return the instrument to its first position and re-centre the plate bubble if necessary, repeating until the bubble is centred in both positions. Swing through 180°, note how much the bubble has moved, and bring it halfway back. Swing through 90° and bring the bubble to the same position with the third foot-screw. The plate bubble should now remain in this position (not necessarily

---

* Most total stations have a so-called compensator, which sets their horizontal circle to be exactly horizontal provided the cup bubble on the instrument (or on the tribrach) lies inside its circular ring. However, the pendulum used to achieve this is susceptible to buffeting when the instrument is used in windy conditions, and this can make it impossible to take a reading. Such instruments therefore have the option to turn the compensator off; if this is done, then the instrument *must* be levelled manually, as described in this paragraph.

† Some instruments have two electronic plate 'bubbles', making it unnecessary to swing through 90°.

central) through whatever angle the instrument is swung. If it does not, repeat the procedure. Levelling in this context really means setting the vertical axis vertical, and should not be confused with the levelling process described in Chapter 6.

9. Check that the instrument is still above the station, and at least to an acceptable degree of accuracy. If it is not, this indicates a bubble error in the cup bubble used for Steps 5 and 6. This can be rectified by moving the instrument on the tripod so that the sighting point is again over the station, and then returning to Step 8.

Note that the alignment of an optical or laser plummet which is built into a tribrach cannot be checked in the manner described in Step 7, even though a misalignment may be present. For this reason, such tribrachs must be regularly serviced and should always be handled with particular care, despite their robust appearance.

4.3.4 Focusing the Telescope

The ability to focus a telescope correctly and thus eliminate parallax in the readings is possibly the single most important factor in obtaining good readings from any conventional surveying instrument, including digital ones. A proper understanding of the next three paragraphs is therefore crucial to almost all surveying.

The objective lens of a telescope creates an image of the distant object on a plane (inside the body of the telescope) whose position can be altered by adjusting an internal focussing lens; the eyepiece has a separate focussing adjustment which allows the observer to view that plane comfortably. To eliminate parallax, it is important that the image of the distant object lies on the plane of the reticle—when it does, the crosshairs will not appear to move against the image of the distant object, even when viewed from slightly different directions. Parallax is therefore eliminated by first using the eyepiece to bring the crosshairs into comfortable focus, and then focussing the objective lens until the image of the distant object is also in comfortable focus; if this is done carefully, the crosshairs and the image should be in the same plane.

Before observing, set the internal focussing lens right out of focus so that no image of the target is visible; and then, with your eye relaxed*, focus the eyepiece so that the reticle is as clear as possible. This adjustment depends only on your eyesight, and not on the distance to any target; once made, it should only need to be altered if your eye starts to get tired.

---

* It is usually easier to relax your eye if you can manage to keep your other eye open, and not tightly shut.

For each observation, adjust the internal focussing lens so that the image of the target is also sharp. This image should then be in the plane of the reticle, on which the eyepiece is focussed. Check this by moving your eye from side to side; there should be no relative movement of the target and the reticle, i.e., no parallax. If there is any, remove it by adjusting the internal focussing lens, *not* the eyepiece. Having done this, it may seem that the target is no longer exactly in focus; in this case, refocus the *eyepiece* so that it is, and check that the parallax has disappeared. If it has not, repeat the procedure above, using the internal focussing lens followed (if necessary) by the eyepiece. Note that parallax can neither be eliminated nor introduced by adjusting the eyepiece—though it may wrongly appear to be present if the eyepiece is badly adjusted for the observer.

4.3.5 Observing the Target

Having eliminated parallax, the crosshairs must be carefully sighted onto the target. Depending on what the target is, it might be easier to bisect it with the single crosshair, or to 'straddle' it with the double crosshairs. If possible, use the same part of the crosshair to sight on each target: this will eliminate any errors due to the crosshairs not being exactly horizontal or vertical. Always use a part of the relevant crosshair (vertical for horizontal angles, and vice versa) which is close to the place where the crosshairs meet.

It is helpful to be aware of the precision with which a sighting must be made, to achieve a given level of repeatability. If a target is 100 metres away, the crosshairs must be sighted onto it with a precision of better than ±1.5 mm, for a reading which is repeatable to within 5 seconds. If the instrument and target are removed and then replaced, it is unlikely that such a level of repeatability could be achieved at all, given the inherent errors of centring over the stations. This understanding is useful when assessing whether a round of observations has 'closed' properly (see Section 4.4.4), or when estimating the accuracy of a horizontal angle for least-squares adjustment (see Chapter 11, Section 11.5). Repeatability can also be affected by changes in light conditions, particularly if the target has a cylindrical or conical shape. If the sun shines on one side of the target, there is a subconscious tendency to sight the crosshairs onto the sunny side, rather than centrally. This error, known as a phase error, disappears when the target is illuminated evenly.

4.3.6 Reading the Angle

In digital instruments with pendulum compensation, this simply consists of reading the numbers on the display. Reading other instruments is a little more complex and is discussed in this subsection.

Every pattern of optical instrument is different in the details of circle reading. Most will have one or more circle-illuminating mirrors, which must be adjusted to obtain even illumination of the circle being observed. Most will have an eyepiece with provision for focussing, through which the circles may be read. In some cases, both circles are visible together, and a single optical vernier applies to both; in other cases, a prism is rotated to select which circle is seen, and separate optical verniers may be used. In each case, rotating the optical vernier causes the scale images to move, by tilting a parallel-sided glass plate in the light path between the circles and the eyepiece. This movement is solely optical and should not be confused with a rotation of the instrument. In higher-precision instruments, the reading system automatically takes the mean of the readings on opposite sides of the circle.

Every time, before reading the vertical circle, level the alidade bubble; this ensures a consistent circle reading when the telescope is truly horizontal. On most modern instruments, this operation is rendered unnecessary by provision of automatic vertical circle indexing (i.e., a pendulum), but without that facility it is *essential* to check the bubble every time when reading vertical angles.

Common errors in reading are to read 10 minutes or a degree out; to read the horizontal circle instead of the vertical circle or vice versa; to fail to level the alidade bubble before reading the vertical circle; or to misread the optical vernier. If you have any doubt on the latter point, it is good practice to turn the vernier to zero, and make your best estimate of what the angle reading should be. Then, turn the vernier to take the reading, watching carefully what happens in the main display as you do so. This always provides the clearest possible indication of how the vernier should be read, in any particular instrument.

4.3.7 Errors Due to Maladjustment of the Instrument

The permanent adjustments to total stations and theodolites are explained in detail in Section 4.5. However it is helpful to outline what these adjustments entail before describing how observations are made, as many steps in this process are directed at eliminating the effects of any maladjustment.

In a properly adjusted instrument:

1. The line of sight of the optical/laser plummet should be along the vertical axis.
2. The trunnion axis should be perpendicular to the vertical axis.
3. The line of sight through the intersection of the crosshairs (called the line of collimation) should be perpendicular to the trunnion axis.
4. The line of collimation should pass through the trunnion and vertical axes.

5. The horizontal and vertical crosshairs should be parallel to the trunnion and vertical axes respectively.
6. When the instrument and alidade bubble are levelled, and the line of collimation is horizontal, the vertical angle should read exactly 0°, 90°, or 270° (depending on how the instrument is configured).
7. For convenience in setting up, the horizontal plate bubble should be central when the vertical axis is vertical.

These points can be checked and permanent adjustments can be made to all of them. (All these adjustments are called 'permanent' to distinguish them from the 'station' adjustments, which are made every time the instrument is set up.)

As well as the requirements listed above, there are others which cannot be adjusted. For example:

8. The horizontal circle should be perpendicular to the vertical axis, and the vertical circle perpendicular to the trunnion axis.
9. The centre of the horizontal circle should coincide with the vertical axis, and the centre of the vertical circle with the trunnion axis.
10. The circles should be accurately graduated.
11. There should be no backlash.

Tests to check the correct setting of instruments may be made, and are described in Section 4.5. If the instrument under test proves to be out of adjustment, refer to the maker's handbook for details. It is unwise to attempt any permanent adjustment to a surveying instrument without prior training.

Although maladjusted instruments can be tiresome to use, most of their effects on observations are eliminated by the observation techniques which are described in the next section.

4.4 MAKING OBSERVATIONS

4.4.1 Principles

It is neither necessary nor possible to ensure that the permanent adjustments described above are always faultless. The effects of these and other instrumental imperfections can be almost eliminated, and careless mistakes can be avoided, by suitable methods of observation.

In measuring horizontal angles, errors due to maladjustment of the trunnion axis are reversed in sign on changing face.* Such errors are therefore eliminated by taking the mean of F1 and F2 measurements.

---

* Transiting the telescope, then rotating the instrument through 180° about its vertical axis.

All collimation errors are also reversed on changing face and are there-fore eliminated in the same way. This applies to both horizontal and verti-cal angles. Since horizontal angles are obtained from the difference of two readings of the instrument, the effect of any error in horizontal collimation is largely (though not totally) eliminated by subtraction. Vertical angles, on the other hand, are measured from a zero in the instrument itself (the alidade bubble or pendulum), and measurements taken on one face only are therefore burdened with the whole vertical collimation error, as well as any error in the alidade bubble or pendulum. It is therefore essential to take F1 and F2 observations for *all* vertical angles, and for all horizontal angles where the full accuracy of the instrument is expected.

Errors due to eccentric mounting and inaccurate graduation of the hori-zontal circle are reduced by repeating the observations (on both faces) using a different part of the circle. This is relatively easy to do on optical instru-ments which (as mentioned above) allow the horizontal circle to be rotated about the vertical axis. On a digital instrument, the only way to physically rotate the horizontal circle is to rotate the entire instrument—either by unclamping the tribrach and rotating it with respect to the top of the tri-pod, or (more quickly but less flexibly) by lifting the instrument out of the tribrach, rotating it by 120°, and putting it back in.

It is common practice to swing the instrument to the right (SR) to observe successive stations when the instrument is face 1 (or CL), and to swing it to the left (SL) when it is in face 2 (or CR). Errors due to any backlash in the instrument are reduced by using the mean of SR and SL measurements, by turning the tangent (slow-motion) screws clockwise for their final adjustment, and by always making any final vernier adjustments in one direction.

Besides eliminating certain instrument errors, taking the average of a number of measurements is desirable in itself; any gross errors in circle reading will be detected and can be discarded. Also, the standard deviation of error (from random causes) of the mean of n measurements varies inversely as $\sqrt{n}$.

All the above precautions are included in the system of observing and booking suggested below.

4.4.2 Practical Points

Before taking any measurements, see that there is no play in the hinges between the legs and the tripod. If the lengths of the tripod legs can be adjusted, see that the clamps are tight. Focus any eyepieces before level-ling the instrument, to reduce the risk of disturbing things later. Start with the tangent screws near the middle of their runs. Once the instrument is levelled, do not jar the tripod or even rest your hands on it. Avoid stepping near the tripod's feet if the ground is soft. Swing the instrument by holding the vertical frames which support the telescope, not the telescope itself. Use the minimum of force on clamping screws.

No routine of observation eliminates errors caused by inaccurate levelling of the instrument, or inaccurate centring over the station mark.

4.4.3 Recording Observations

Most instruments which display readings digitally are also capable of recording them onto a memory card at the touch of a button. (Some of these instruments can also take a photograph at the same time, to provide a definitive record of what has just been observed.) The remainder of this subsection covers cases where the readings are to be recorded manually.

Haphazard observation and random booking on loose paper lead to mistakes. For speed and accuracy, a system is essential. The one given here is the result of more than a century of experience in combating human and instrumental error.

For efficient work using an optical instrument, a separate observer and booker are necessary. With a digital instrument it is possible for one person to observe and do the booking as well, as there is less arithmetic to be done in the field.

Record all necessary data in an observation book. Use a fresh page for each station occupied. Book with a ballpoint pen or pencil, and make your figures neat and clear. *Do not erase*; make corrections by drawing a single line through the incorrect figures, leaving them legible, and writing the correct figures beside or above them. A fair copy may be made later on another page if necessary (check carefully for copying errors), but *the original pages must not be discarded*. Notice that single-figure entries are written 06° 08′ 05″, not 6° 8′ 5″

The booker fills in the heading and the stations to be observed while the observer is setting up the instrument. The observer calls out the readings; the booker records them and then reads back what (s)he has written; the observer then rechecks the reading and replies 'correct' (or not). Do not omit this seemingly pedantic precaution—it ensures that there is a 'closed loop' between what is visible in the instrument and what is written in the book.

It is the booker's duty to detect inconsistencies (such as excessive discrepancy between F1 and F2 readings); if one occurs, the observer is at once told to check, but is not told what is wrong. In the case of an optical instrument, the booker works out reduced angles[*] while the observer is observing the next target. After a bit of practice, *mental arithmetic is both quicker*

---

[*] Note that this is *not* the same as the booker repeating what he or she has just heard!

*and less error-prone than use of a calculator in the field.** The booker is also responsible for ensuring that the stations are observed in the right order. Both surveyors should do their own jobs and not interfere with the other. Ideally, the observer checks the booker's arithmetic, and both initial the sheet before leaving the station.

Remember that in practice, the time taken in going to and from a station is large compared with the time actually spent there. So take every reasonable precaution to ensure that carelessness does not make a second visit necessary.

4.4.4 Horizontal Angles

Take at least one full round of observations, i.e., F1/SR *and* F2/SL. In the F1 position, sight onto the reference object (R.O.) and, on a digital instrument, set the horizontal angle to zero. On an optical instrument, rotate the horizontal circle so as to show a random value just over 0°, as this will simplify the subsequent arithmetic. Do not attempt to set it up to show an exact angle, but set it reasonably close to the value you want, and then use the vernier to measure exactly what you have achieved.

Record the reading then, and by swinging right throughout, take and record readings for all the other points to be observed from the station. Always close the round by re-observing the R.O.; the difference in the two readings is called the closing error. A standard for acceptable closing errors will depend on the instrument in use and on the quality of the work required—a typical value might be 5 seconds.† If there is an unacceptable closing error, all the readings should be discarded and a fresh start made.

Change face by transiting the telescope and swinging through 180°. Sight back on to the R.O. again, and set the horizontal angle to zero again if using a digital instrument. On an optical instrument, change the position of the horizontal circle just slightly—ideally by about half the travel of the vernier scale. (This evens out any systematic inaccuracies in the vernier scale and also guards against repeated misreading of the scales or unconscious memories of previous readings.)

Record the new reading to the R.O. (0° for a digital instrument, or just over 180° for an optical one) and, swinging now from right to left, take the readings for all the points to be observed from the station, in the order of reaching them—i.e., the opposite order from the F1 readings. Close the half-round again as before.

---

* A calculator can be used when the booking sheet is checked, especially if this is done in the office—doing so provides a fully independent check of the arithmetic done in the field.

† Careful selection of a reference object can help in obtaining a small closing error. If the object is too far away, atmospheric distortions can impair the repeatability of the observation; if it is too close, the size of the target in the eyepiece will make it hard to sight the instrument in exactly the same way. Remember that 1 second represents a distance of 1 mm at a range of 200 metres.

HORIZONTAL/~~VERTICAL~~ OBSVNS AT: Station Y GROUP: A2 PAGE
Date: 5/4/2013 Instrument: TC405 no. 3 Observer: A. Smith JOB: Major
Time: 15:45 Ht of inst: Booker: B. Jones Control
Weather: Overcast Checker:
Temperature (°C): Pressure (mm Hg):

Face/ swing	Stations and points	Observed Angle °	'	"	Mean F1/F2 °	'	"	Raw/Corrected HD/SD (delete as necessary)
F1	Station Z (R.O.)	00	00	00				
SR	Church spire	42	43	22	42	43	26	
	Chimney l/c	166	36	15	166	36	23	
	Station X	207	~~42~~ 24	11	207	24	13	
	R. O. (check)	359	59	57				
					360° – VA			Target height
F2	R.O. (check)	00	00	04				
SL	Church spire	42	43	29				
	Chimney l/c	166	36	28				
	Station X	207	24	15				
	Station Z (R. O.)	00	00	00				

Figure 4.3 Booking horizontal angles (digital instrument).

When booking, it is always best to write the points for SL in the same order as for SR, and to book from the bottom of the form upwards on SL. After the first half round, and throughout a repeated round, the booker is in a position know in advance what the next reading should be, and should ask the observer to check if there is an unacceptable discrepancy.

When a full round of horizontal angle observations has been completed, the booker should enter the mean of the F1 and F2 angles (reduced, if necessary—see Figure 4.4) in the top part of the right-hand column. The bottom part of the same column can be used to record any horizontal or slope distances which are observed. Figure 4.3 shows a competed booking sheet for a digital instrument, while Figure 4.4 shows the same round observed by an optical instrument. Note that instrument heights are only needed when vertical angles or slope distances are measured, and that temperatures and pressures are only required when 'raw' distances are recorded, for subsequent correction in the office.

4.4.5 Vertical Angles

For vertical angles, *always* take sets of observations both on F1 and on F2 (see Figure 4.5). There is no virtue in swinging left or right, and there is no R.O. to close on. The horizontal crosshair may not be quite horizontal, so always intersect with the same part of it, just to one side of the vertical

HORIZONTAL/~~VERTICAL~~ OBSVNS AT: ___Station Y___   GROUP: _A2_   PAGE
Date: _5/4/2013_ Instrument: _T2 no. 6_ Observer: _A. Smith_   JOB: _Major_
Time: __15:45__   Ht of inst:_____   Booker: _B. Jones_   _control_
Weather: __Overcast__   Checker:_____

Circle/ swing	Stations and points	Observed Angle ° ' "			Reduced/Angle ° ' "			Mean CL/CR ° ' "		
CL	Station Z (R.O.)	00	08	15	00	00	00			
SR	Church spire	42	52	37	42	43	22	42	43	26
	Chimney l/c	166	44	30	166	36	15	166	36	21
	Station X	207	32 ~~12~~	26	207	24	11	207	24	13
	R. O. (check)	00	08	12	359	59	57			
								HD/SD/Target Height		
CR	R.O. (check)	180	05	43	00	00	04			
SL	Church spire	222	49	08	42	43	29			
	Chimney l/c	346	42	07	166	36	28			
	Station X	27	29	54	207	24	15			
	Station Z (R.O.)	180	05	39	00	00	00			

Figure 4.4 Booking horizontal angles (optical instrument).

~~HORIZONTAL~~/VERTICAL OBSVNS AT: ___Station Y___   GROUP: _A2_   PAGE
Date: _5/4/2013_ Instrument: _TC405 no. 3_ Observer: _A. Smith_ JOB: _Point_
Time: __17:05__   Ht of inst: _1.465 m_   Booker: _B. Jones_   _fixing_
Weather: __Sunny periods__   Checker:_____
Temperature (°C):_____ Pressure (mm Hg):_____

Face/ swing	Stations and points	Observed Angle ° ' "			Mean F1/F2 ° ' "			~~Raw/Corrected HD/SD~~ (delete as necessary)
F1	Station Z	90	08	35	90	08	42	14"
	Marker 1	85	52	00	85	52	08	17"
	Marker 2	78	44	30	78	44	36	12"
	Station X	87	30	25	87	30	32	15"
					360° – VA			Target Height
F2	Station Z	269	51	11	90	08	49	1.423
	Marker 1	274	07	43	85	52	17	
	Marker 2	281	15	18	78	44	42	
	Station X	272	29	20	87	30	40	1.508

Figure 4.5 Booking vertical (zenith) angles.

hair—remember that the left-hand side of the reticle on F1 becomes the right-hand side on F2.

The booker records the height of the instrument's trunnion axis above the station mark—note that, without this measurement, all vertical angle observations will be useless! If a target on a tripod is being observed, the height of the target above the station will also be needed, and can be recorded in the lower part of the right-hand column. If some nonstandard target is being observed (e.g., part of a weathervane), the observer should make a sketch in the observation book, indicating by an arrow the exact point intersected on the object.

Assuming that the instrument measures zenith angles, the horizontal will appear as 90° on F1 and 270° on F2. Next to the F2 readings, the booker should record the value obtained by subtracting each reading from 360°. The differences between these and the F1 observations can be booked in the upper part of the right-hand column, and should remain nearly constant; the constant is zero only if the permanent adjustment on the vertical circle is perfect. If any difference varies significantly from the norm, the booker should demand a check.

Finally, the column beside the F1 observations should be used to record the accepted vertical angle for each observation, namely the average of F1 and (360° − F2). This exactly cancels out any maladjustment of the vertical circle.

If the instrument measures slope angles, then there is no need to subtract 360° from the F2 readings; otherwise, the procedure is exactly as described above.

If the instrument does not have a compensator, *always centre the alidade bubble* before taking each reading of the vertical circle.

4.4.6 Setting Out Angles

As mentioned in Chapter 3, it is often necessary to set up an instrument to sight in a predetermined direction, rather than simply to record the direction it is sighted in when observing a target. Usually the starting point is a known horizontal angle which must be swung through, once a reference object has been observed.

With a digital instrument, the process is quite simple. A booking sheet can be filled out beforehand with the stations and observed angles (except for the check on the R.O.), as a convenient way of carrying the data out to the field. Sight on the reference object with the instrument in face 1, and set the horizontal angle to zero. Then swing the instrument to approximately the right direction, clamp it if necessary, and turn the horizontal tangent screw until the required horizontal angle is displayed. Markers can then be set up at any point along the line of sight. Finally, the reference object is sighted again (and its angle recorded), to ensure that no setting has been disturbed. For accurate work, this process is then repeated with the instrument in F2, and an average of the two sight lines is used.

With an optical instrument, a booking sheet such as the one in Figure 4.4 is partially filled in before going out into the field, with the 'Circle/swing', 'Station,' and 'Reduced Angle' columns all completed using the known data. Once out in the field, the reference object is observed in circle left, and the observed angle is recorded in the usual way. The reduced angle of the required direction(s) is then added to this reading, and written down in the 'Observed Angle' column against the point(s) to be set out. The observer is told to swing the instrument to this angle, and does so by first setting the optical vernier to read the appropriate number of minutes and seconds, then swinging the instrument and using the horizontal tangent screw until the entire correct reading is shown in the display.[*] As with the digital instrument, the reference object is then re-observed (and booked), and the process repeated in circle right.

See Appendix F for a worked example in calculating an angle for setting out, including the preparation of a booking sheet for a digital instrument.

4.5 CHECKS ON PERMANENT ADJUSTMENTS

It is usually impossible to make corrections to permanent adjustments in the field—and often unwise to attempt them back at base either, unless clear instructions are given in the instrument manual. However, it is necessary at least to know when an instrument needs to be professionally serviced, and to give the correct diagnosis of the problem. The following guidelines should assist in this; the simpler adjustments are described first.

4.5.1 Bubble Errors

If there is a significant error in the plate bubble when levelling the instrument (see Section 4.3.3 above on setting up, Step 8), this can be removed by first levelling the instrument carefully, then adjusting the plate bubble so that it lies in the centre of its glass. In a level instrument with no plate bubble error, the instrument can be rotated about its vertical axis and the bubble will always return to the centre point.

If the instrument has been levelled using the plate bubble, and the cup bubble on the tribrach is no longer central, then this indicates a bubble error on the tribrach. Again, this can be removed by levelling the instrument using its plate bubble, and then adjusting the mounts for the tribrach bubble until it too is centred.

---

[*] The horizontal angle between each station and the reference object. This is displayed directly on a digital instrument, but must be calculated (by subtracting the reading for the reference object from the reading for each successive station) when an optical instrument is used.

If there is a significant and consistent difference between the F1 and (360° − F2) readings on zenith angle observations, this indicates a bubble error in the alidade bubble. Putting this right is more complex. The procedure is to make a vertical observation, and calculate the average of the F1 and (360° − F2) readings. With the instrument observing the target in F1, adjust the alidade bubble until, with the bubble central, the average reading is obtained. Then re-observe on F2 and check that (360° − F1) gives the same angle.

4.5.2 Plummet Errors

If the plummet (laser or optical) rotates with the instrument, it is easy to see whether there is a plummet error by simply rotating the instrument. If the line of collimation makes a circle, then there is an error; this can be removed by keeping the instrument still, and adjusting the line of collimation of the plummet to point at the centre of the circle. Details of how to do this should be given in the instrument's manual.

If the plummet does not rotate (e.g., it is fixed to a tribrach), then errors are harder to detect. One simple method is periodically to lay an instrument on its side on a bench, with the tribrach attached and the vertical axis clamped. Secure the instrument from rocking, sight through the plummet, and mark the point on the wall* that lies on the line of collimation. Unclamp the vertical axis, rotate the tribrach through approximately 120°, and repeat; then rotate through a further 120°, and repeat again. If the three marks on the wall are in different places, the plummet has an error. Correcting such an error generally requires a special instrument or professional servicing.

4.5.3 Reticle Errors

Reticle errors should only be corrected by an instrument maker, but are relatively easy to diagnose. To check the vertical reticle, sight on a suitable target with some part of the reticle, then rotate the telescope about the trunnion axis and see whether all parts of the vertical reticle align with the target. The horizontal reticle is similarly checked, by rotating the telescope about the vertical axis. These tests show whether the horizontal and vertical crosshairs are at right angles to the vertical and trunnion axes respectively, but not whether the two crosshairs (or the two axes) are at right angles to each other.

---

* There is no need to make a mark on the wall itself—a piece of paper can be fixed to the wall with masking tape.

4.5.4 Collimation Errors

Collimation errors in the horizontal plane can be detected by sighting on a target at a similar height to the instrument and taking a horizontal angle reading; then transit the instrument, and take the reading again. If the two readings do not differ by exactly 180°, then this is probably due to collimation error. The target should be as close as possible to the instrument (consistent with being able to focus on it, and repeat each reading reliably), as this magnifies the effect of any collimation error. As explained above, this error is relatively unimportant since it partially cancels when one horizontal observation is subtracted from another, provided the two targets are at a similar distance from the instrument—and it cancels further when the mean of F1 and F2 readings is used.

Collimation errors in the vertical plane have no real meaning in instruments with an alidade bubble: any such error is effectively treated as a bubble error. In pendulum instruments, any discrepancy between F1 and (360° − F2) zenith angle readings is likewise compensated for by adjusting the pendulum.

4.5.5 Trunnion Axis Misalignment

If the F1 and F2 horizontal angles between two objects are seen to differ when the two objects are at similar distances from the instrument but at different heights, the most likely cause is misalignment of the trunnion axis. The simplest way to check for this is the so-called spire test, which involves setting up the instrument where it can observe a well-defined target along a steeply sloping line of sight—e.g., a church spire. Sight onto the target, then rotate the telescope about the horizontal axis until it is sighting onto a patch of ground about 10 or 15 metres in front of the instrument. Check that it is possible to focus the telescope (sight a bit further away, if not), lay a ruler or tape measure on the same patch of ground such that it is parallel with the horizontal crosshair, and record the reading (in millimetres) where the vertical crosshair intersects it. Change face, sight on the target again, swing the telescope down, and see where the vertical crosshair now crosses the ruler. A difference of 1 mm in the two readings suggests (using reasonable estimations of the likely geometry) that the trunnion axis is misaligned by about 10 seconds. On a 1-second instrument, a misalignment of this magnitude should be corrected—usually by the manufacturer.

Chapter 5

Distance Measurement

5.1 GENERAL

Distances may be measured by five methods: tape, optical, electromagnetic, ultrasonic or GNSS. The method used on any particular job depends upon the number and size of the distances to be measured, the nature of the ground, the accuracy required, the time available, and the availability of suitable equipment. For all but the smallest or largest tasks, electromagnetic distance measurement (EDM) is the simplest choice for terrestrial measurement. GNSS measurements provide interpoint distances over any distance and without the need for intervisibility, but the procedure is rather more complex.

Before making any measurement, it is always wise to obtain an estimate of its value by an approximate method, to reduce the possibility of gross errors. At least three methods are available for this:

1. Pacing is used for rough measurements and to check accurate measurements against gross errors. Test your natural pace over a measured distance, rather than trying to pace metres. The accuracy on smooth ground is about 1 part in 50.
2. A perambulator is a wheel fitted with a revolution counter, and is wheeled along the line to be measured. It is more accurate than pacing, and is frequently used in measurement for costing of highway repairs.
3. If a suitable map is available, scaling from it will give a close approximation.

5.2 TAPE MEASUREMENTS

Tapes are now mainly used for the quick measurements of short distances (horizontal or vertical). However, they used to be the most accurate method of measuring all distances, so their use was developed to a fine art by surveyors in the first part of the 20th century. This involved suspending the tape from a tripod at each end under a known tension (such that it formed a catenary), and reading the point where the tape passed over each tripod using a micrometer. For further details, see Bannister, Raymond and Baker (1998).

Tape measurements are subject to the following sources of error:

1. Inaccuracy in the markings on the tape
2. Variations in the length of the tape due to changes in temperature
3. Variations in the length of the tape due to changes in tension
4. Slope (since it is usually the horizontal component of the length that is required)
5. Sag on any unsupported spans, if this is not allowed for in the calculations
6. Errors at the junction of tape lengths

Fabric tapes made of linen or (preferably) fibreglass are used for low-accuracy or detail work. Steel bands or tapes are more accurate but are easily damaged if kinked or trodden on. On smooth ground an accuracy of about 1/2000 is attainable. For the highest accuracy:

1. Calibrate the tape against a known distance, at the same temperature and tension as will be used on the job.
2. Avoid large changes of temperature by working early, late, or on a cloudy day.
3. Use a steady pull, ideally by means of a spring balance.
4. Correct for slope: if there are marked changes of gradient, measure the gradient and note the (slope) length of each section.
5. Use the longest tape possible, if the distance is greater than one tape length.
6. Take the mean of two measurements in opposite directions.

5.3 OPTICAL METHODS (TACHYMETRY)

The stadia lines (see Figure 4.1) on the reticle of an instrument's telescope subtend a particular angle α, usually 0.01 radians. Thus, if a distant vertical staff is viewed horizontally by means of (say) a level, the distance D from the instrument to the staff is given by s/α (i.e., usually 100 s) where s is the length of the staff between the two stadia lines. Either an ordinary levelling staff or a specially-made tachymetry staff may be used.

This method of determining distance is perfectly straightforward when the line of sight is horizontal—and modern digital levels also use this principle to measure the distance to their (bar-coded) levelling staves. However, it becomes more complex if the line of sight needs to be inclined by some angle to the horizontal in order to observe the staff. It is not practicable to hold the distant staff perpendicular to such a line of sight; instead, it is still held vertical by means of a bubble, and the two readings (plus the vertical angle of the telescope) can be used to obtain both the horizontal distance to the staff, and the height difference between the instrument and the base of the staff.

The calculations for this form of tachymetry are tiresome, and the technique is now obsolete[*]. In any case, the accuracy of vertical staff tachymetry is always limited by the fact that lines of sight defined by the two stadia hairs are differently affected by atmospheric refraction.

For distances up to about 50 m, a more accurate form of horizontal staff tachymetry, known as subtense, is still occasionally used. A special staff called a subtense bar, usually 2 m or 3 m long, is mounted horizontally and at right angles to the instrument's direction of view. The horizontal angle subtended at the instrument is then measured, and the slope distance between the instrument and the staff can be deduced by simple trigonometry.

5.4 ELECTROMAGNETIC DISTANCE MEASUREMENT (EDM)

5.4.1 Principles

The principle of the method depends on measuring the transit time of an electromagnetic wave which is transmitted along the line and reflected back to the transmitter. Some devices such as LiDAR laser scanners transmit a pulsed laser beam, and simply measure the time taken for the pulse to be reflected—this can be done without the need for a special reflector at the far end of the line, and so is sometimes known as a 'reflectorless' system. Others use a carrier wave, modulated at a known frequency, and measure the phase change of the reflected modulation to calculate the distance (see Figure 5.1). Errors can arise from difficulties in knowing the exact point of measurement within the instrument (a matter of a few millimetres), from inaccuracies in measurement (about 2 or 3 parts per million for a modulating device) and from variations in air density along the path of the wave (up to about 20 parts per million, if no correction is made).

The reflectors (for those instruments which require them) take the form of 'corner cubes' with precisely ground faces, so that incoming radiation is reflected back in the exact direction that it came from. It is important to use only the correct type of reflector with a given instrument, since each type has a different 'reflector constant', depending on the path length of

---

[*] The formulae, if needed, are given in Uren and Price (3rd ed., 1994), p. 144.

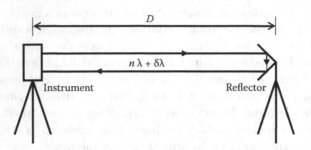

Figure 5.1 Electromagnetic distance measurement (modulated carrier wave).

the optics and the density of the glass.* Neglect of this factor will lead to a systematic error in all measurements.

Most modulating EDM devices now use infrared or a visible laser (wavelength of a few microns) as the carrier wave, modulated at around 1 GHz to give a modulation wavelength of about 30 cm. Their range can be up to 30 km, though measurements of more than about 5 km are now generally done using GNSS. Some earlier instruments used microwaves (wavelength of a few cm) as a carrier, and were capable of measuring up to 100 km, provided that the end stations were intervisible.

The total distance travelled by the modulated wave, $2D$, will be equal to a number of whole wavelengths $n\lambda$ plus a fraction of a wavelength $\delta\lambda$ (see Figure 5.1). δ is relatively easy to determine by comparing the phase of the reflected wave with that of the transmitted wave. Some possible methods are:

1. To use a phase discriminator circuit to compare phases directly.
2. To shift the phase of the reflected signal by a known amount until it gives a null with the reference signal.
3. To use a digital count of time signals between the reference null and the reflected null, then multiply by the modulation frequency to find the phase difference.

The value of n can be found by increasing the modulation frequency by a small fraction, and measuring the change in δ. Typically, the modulation frequency is increased by 1%; the resulting fractional change in δ is then multiplied by 100 and rounded down to the nearest integer to give the value of n. The change in δ is measured by changing to the new frequency and subtracting the new value of δ from the old value, adding one whole cycle in cases where δ appears to have decreased. However, this method only works properly when n is less than 100: any multiples of 100 would pass

---

* Most modern total stations allow their reflector constant to be changed, so they can work with different reflectors—but it is important that this constant is correctly set for the reflector that is being used.

undetected. Such multiples can, however, be counted by altering the frequency by 0.01% (1% of 1%) and again measuring the fractional change in δ. Long-range machines may thus need to make several changes of modulation frequency to compute the distance properly.

5.4.2 Use of EDM

For short-range work, hand-held laser devices can be used to measure distances with an accuracy of around 3 mm, without the need for a reflecting target at the far end of the ray. These devices measure to the exact point which is illuminated by the laser measurement system; the more sophisticated ones include an on-board camera and liquid crystal display to show the user which point is being illuminated, when this is too far away to be seen by the naked eye.

EDM systems mounted on tripods (typically in the form of total stations) are used for high accuracy (2 or 3 parts per million) measurements of distances between, say, 10 m and 5 km. (Longer distances are now generally measured using differential GNSS, as described in Section 5.6.)

The basic measurement made by an EDM device is a slope distance between the instrument and target, uncorrected for atmospheric conditions. On most instruments, it is possible to correct for atmospheric conditions while in the field. Temperature and pressure are measured near the instrument and, for accurate work, near the target. On some instruments these readings are typed in and the instrument uses a built-in formula to calculate the correction; on others, a nomogram or chart is provided to convert these readings into a parts-per-million correction, which is then entered into the instrument. To guard against undetected errors, it is wise also to record the temperature, pressure and the uncorrected distance, so that the field correction can be checked back in the office.

Some EDMs do not always show the whole of the distance that they are measuring; for instance, the display might show 2,345.678, when the distance is actually 12,345.678. A 10 kilometre error should not easily pass unnoticed; but a useful field check is to record the readings obtained with the atmospheric correction set first to +50 parts per million, and then to −50 ppm. The difference between the two readings, when multiplied by 10^4, gives the approximate distance—and any 'overflow' in the display will then be detected quite easily.

In addition to slope distance, most EDMs are also able to calculate the horizontal and vertical distance between the instrument and target. In the case of horizontal distance, this has the advantage that the heights of the instrument and target above their stations do not need to be measured, since the horizontal distance between the instrument and target will also be the horizontal distance between their respective stations. (For the other distance measurements to be useful, it is vital that these measurements *are* recorded.)

To make a calculation of horizontal or vertical distance, the micropro-
cessor inside the EDM must know the vertical angle between the instrument and target. This means that the instrument must be carefully aimed at
the target—it is not sufficient simply to sight it well enough for the radiated
signal to be returned. If necessary, the alidade bubble on the instrument
should also be adjusted to show the correct vertical angle.

For precise or long-distance work, however, these calculated distances
must be treated with caution. A fully accurate 'horizontal distance' calculation involves knowing the height of the instrument above sea level; and
both calculations have to make some assumption about how light curves
in the atmosphere, which may not be valid at the time of the measurement.
Chapter 10 discusses both of these issues in detail.

5.5 ULTRASONIC METHODS

Ultrasonic distance measuring devices work in a similar way to reflectorless EDM devices; namely, by transmitting a pulse of very high frequency
sound and measuring the time taken for it to be reflected. They are generally cheaper than hand-held EDM devices, but they have a shorter range
(typically less than 20 metres) and are less accurate (about 1 part in 200).
They often incorporate a laser beam, which can make them appear to be an
EDM device—but the purpose of the laser is simply to show the operator
where the device is pointing when the measurement is taken. However, the
sound pulse has a much wider beam angle than the laser, so the distance
measurement can be erroneous if it is reflected more strongly from a surface
other than the one which the laser is illuminating.

The accuracy and reliability of ultrasonic devices depends on the hard-
ness of the surface to which the distance is being measured, rather than its
light-reflecting properties; this makes ultrasonic devices less effective than
EDMs for measuring distance to soft surfaces such as foam rubber, but better than EDMs for measuring the distance to transparent surfaces such as
a sheet of glass or the surface of a clear liquid. Ultrasonic devices thus have
some valuable applications, but their lack of range and accuracy mean that
they cannot properly be considered as a tool for surveying.

5.6 GNSS

As explained in Chapter 7, differential GNSS (two instruments in different
locations, receiving signals from the same satellites simultaneously) is able
to find the relative positions of the two locations to an accuracy of about
2 mm per kilometre of separation, i.e., an accuracy of 2 parts per million.
This makes DGNSS a useful distance measuring tool, particularly as it does

not require a line of sight between the endpoints. Moreover, there is no need to apply any co-ordinate transformation to the results—the fundamental positions reported by GNSS receivers are Cartesian co-ordinates as defined by WGS84, and the slope distance can be found by simply computing

$$\sqrt{(x_2 - x_1)^2 + (y_2 - y_1)^2 + (z_2 - z_1)^2},$$

where the subscripts refer to the two receiver locations.

Although the instruments required for DGNSS are more expensive, they are slightly more accurate than EDM and they save time, especially in rough country and in conditions where the line being measured is obstructed (e.g., by buildings) or is subject to interference by traffic.

Chapter 6

Levelling

Surveyors frequently need to find the relative heights of two or more control stations. There are various ways of doing this, including GNSS techniques (Chapter 7) and trigonometric heighting (Chapter 12), but for points which are close to each other, the simplest and most accurate process is called levelling.

The method involves an instrument called a level, and a staff. A level is a telescope mounted on a tripod, with some means for setting the line of collimation (the sighting line) to be exactly horizontal. A staff is simply a long ruler; an optical staff is usually graduated in centimetres and a digital staff has a long bar code which can be read by a digital level. Most digital staves* also have a conventional (i.e., optical) scale on the other side of the staff.

6.1 THEORY

To find the difference in level between two points A and B (see Figures 6.1 and 6.2), the observer sets up the instrument at an arbitrary third point I_1. An assistant holds the staff vertical with its foot resting on A. The observer rotates the telescope about its vertical axis until the staff appears in the centre of the field of view, sets the line of collimation to be horizontal, and then reads the scale of the staff against the horizontal crosshair (distance a in Figure 6.1). The staff is then moved to B and the observer again directs the telescope onto it, obtaining reading b. The difference in level between A and B is then $(a - b)$, since, if the instrument is correctly adjusted, both lines of collimation are horizontal. The height of the instrument at I_1 does not affect the calculation.

If the height of a third point C, beyond B, is required, the instrument is moved to I_2, between B and C. The difference in level between B and C is then found in the same way. By repeating this process, the difference in level between points at any distance apart can be found.

---

* The plural of staff.

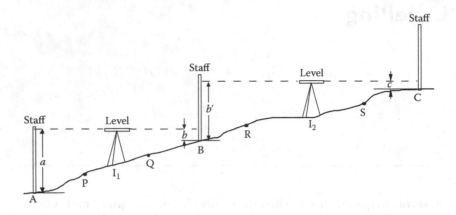

Figure 6.1 Measurement of height differences using a level.

In levelling from a mark at A whose level is known (e.g., a site datum or other benchmark), the observation I_1A is called a backsight. This establishes the height of the instrument (called the line of collimation) at I_1. The observation I_1B is called a foresight. Similarly I_2B is a backsight, I_2C is a foresight and so on.

A *line* of levelling is usually started at one benchmark of known height and, if possible, finished at a different one. A long line should be broken into in a series of *bays*, of between one and five instrument positions. Each bay should be 'closed' by being levelled out and then back to its starting point, as a check against error. Each bay should run between two well-defined markers which can support the staff and which will not change in height: either a permanent benchmark such as those provided by the Ordnance Survey in the UK, or a temporary one, such as a stout peg. It is helpful if the point on which the staff stands is convex, so that there is a uniquely defined 'highest point' on it, which is taken as its height.

Figure 6.2 shows a line of levelling in plan view. The first bay runs from a benchmark at A out to C (via B) and then back to A (via D); points B and D are called change points, and point C is a temporary benchmark or TBM. A second, smaller bay then runs from C to E and then back to C. Finally an 'open' bay is run from E to Z; if this gives a height for the closing benchmark which is in good agreement with its published height, then there is no particular need to close that bay back to its starting point, and the calculated heights of all points in the line can be accepted. If the agreement is not good, then the bay *should* be closed; if it closes well, then it raises the possibility that the first or last benchmark might have subsided since its height was last checked.

If the heights of further points are required, the staff is held (say) at P and Q and at R and S, and readings are taken from I_1 and I_2 respectively. If

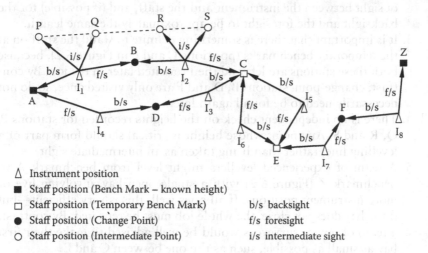

△ Instrument position
■ Staff position (Bench Mark – known height)
□ Staff position (Temporary Bench Mark)   b/s backsight
● Staff position (Change Point)   f/s foresight
○ Staff position (Intermediate Point)   i/s intermediate sight

Figure 6.2 Typical line of levelling (plan view).

these intermediate points lie in a straight line, and the horizontal distances between them are measured, a vertical section of the ground (called a level-section) can be plotted.

Another use for intermediate points arises when a given height needs to be set out, typically on a vertical wooden post driven into the ground. The level is sighted at the post, and a pencil mark is made at the height of the line of collimation, i.e., where the horizontal crosshair intersects the post. The height of this mark can be calculated by simple arithmetic, and a tape measure can then be used to measure up or down the post to the required height level.

Observations to intermediate points are called intermediate sights. They can be made more quickly than sightings to change points, because several sightings can be made from each instrument position; but they are also more prone to undetected errors, because they do not form part of a closed bay.

Note the following:

1. When the staff is moved, the instrument must remain stationary; and when the staff remains stationary, the instrument *must* move, to guard against a reading error going undetected. (If the instrument was not moved from I_5 to I_6 in the second bay, then a misreading to the staff at E would 'cancel out' to give a bay which appeared to close well, yet gave the wrong height for E.) The only times when the instrument and staff both move are at the start of a new bay or at the end of the job.

2. The instrument positions I_1, I_2, etc. need not be on the straight lines between AB and BC, etc. The only requirement is to have a clear line

of sight between the instrument and the staff, and (if possible) for the backsight and the foresight to be approximately the same length.

3. It is important that there is something definite to stand the staff on at the temporary benchmarks (positions C and E in Figure 6.2), because both these stations are left and then revisited later in the job. By contrast, change-point stations B, D and F are only visited once, so do not necessarily need to be found again later.

4. There is no independent check on the heights recorded for stations P, Q, R and S. Any point whose height is critical should form part of a levelling line, rather than being taken as an intermediate sight.

5. A team of experienced levellers might level from benchmark A to benchmark Z (Figure 6.2) with a single open bay, involving five or more instrument positions. If all goes well, this is very efficient—but if the bay does not close, the whole job must be repeated. By contrast, a team of novice surveyors would be well advised to make their first bay as small as possible, such as the one between C and E.

6.2 THE INSTRUMENT

A basic optical level consists of a telescope with a reticle similar to that in a total station (see Figure 4.1), which can be rotated about a vertical axis. The instrument has a cup bubble which is first used to set the vertical axis approximately vertical, a tangent screw to aim the telescope, and a more sensitive bubble to allow the line of sight to be set exactly horizontal once the telescope has been sighted on the staff. On such instruments, it is necessary to adjust the instrument using the sensitive bubble each time the telescope is sighted in a new direction.

The stadia lines on the reticle allow the instrument to be used for measuring distances approximately. The two lines typically subtend an angle of 0.01 radians, which means that every centimetre of difference between the upper and lower readings implies a metre of distance between the instrument and the staff.

In a self-setting or automatic level, a stabiliser or pendulum automatically levels the line of collimation for every sighting. The stabiliser consists of one prism fixed internally to the telescope casing and two prisms which are suspended freely as a pendulum within the telescope. When setting up, a cup bubble is first centred to ensure that rotational axis of the telescope axis is not more than about ±20′ from the vertical. The pendulum is then automatically released from its clamps and the staff can be read at once. This type of instrument saves much time when running a line of levels. It is, however, more subject to interference in windy conditions.

A digital level typically has a stabiliser and can be used to read a conventional staff in the same way as an optical level. In addition, though, it has the capacity to read a digital staff, usually to a higher degree of accuracy, and display the results (height reading and distance to the staff) on a screen. More accurate digital levels will take a reading several times and display the average (and standard deviation) of the results; the observer should be prepared to reject a set of readings if the standard deviation is too high, as this implies that the staff was not being held absolutely steady. Most digital levels also have the ability to record their readings electronically.

Digital levels differ from optical levels in that they do not just read the part of the staff which lies behind the horizontal crosshair, but scan the staff a short distance above and below that point. This means that a digital level is unable to read to the very top (or very bottom) of a staff. When planning a reading, it is therefore good practice to ensure that neither the top nor the bottom of the staff is visible in the field of view of the telescope. It is also important to focus the telescope carefully on the staff and to eliminate parallax (see Chapter 4, Section 4.3.4), as this will affect the accuracy of the digital readings.

In a tilting level, the telescope and the sensitive bubble are pivoted on a horizontal axis, and can be slightly elevated or depressed by means of a micrometer screw. The instrument can be used as described above for levelling—but predetermined slopes can also be set out, using a graduated ring mounted on the drum of the micrometer screw.

On a basic optical level, the staff is read by estimating the reading which lies exactly behind the horizontal crosshair. Since most optical staves are graduated in centimetres, this makes it hard to measure a height to better than about 2 mm. To obtain higher accuracy, optical levels may incorporate a device known as a parallel plate micrometer—a thick disc of glass with a high refractive index and surfaces which are exactly parallel, which lies on the light path between the main telescope and the distant object. This disc pivots about a horizontal axis which is parallel to its two circular surfaces and perpendicular to the telescope's line of collimation. When the two surfaces of the disc are vertical, this has no effect on the view through the telescope—but as the disc is rotated (by means of a micrometer), the image of the staff appears to move up and down against the horizontal crosshair, as shown in Figure 6.3. The observer rotates the disc until one of the height markers on the staff coincides with the horizontal crosshair, and then reads the graduated micrometer to see how much additional height should be added to the value observed on the staff.

At normal ranges (up to about 50 m), simple optical levels permit readings to be estimated quite easily to within 5 mm, while standard digital levels are typically accurate to about ±1 mm. For work requiring greater accuracy, precise digital levels or optical levels with parallel plate micrometers allow readings to be taken to submillimetre accuracy, and even down to 0.01 mm. Levelling to this degree of accuracy is called 'precise level-

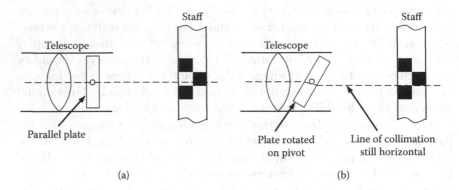

Figure 6.3 Parallel plate micrometer.

ling' and involves special staves and techniques which are fully described in Section 6.6.

Measurements of distance (whether using the stadia lines on an optical instrument or a digital readout) rely on a method called tachymetry (see Chapter 5, Section 5.3) and are not particularly accurate; if the stadia lines are being read to ±5 mm, the distance will only be accurate to about 1 metre at best. The particular system in each instrument is simply designed to be accurate enough for its intended purpose—namely, to keep the distances of foresights and backsights approximately equal, as discussed below.

6.3 TECHNIQUE

Most modern levels (especially precise levels) read the staff digitally; some of the guidance below is only for optical levels, but most is relevant to both types.

Set up, focus and eliminate parallax as with a total station (see Chapter 4, Section 4.3.4)—this is vital for accurate readings, even with a digital instrument. Level the instrument if necessary, as described above. Never rest your hands on the tripod while observing.

We have assumed that the line of collimation is horizontal when the instrument is levelled. This is only true if the instrument is in exact adjustment. If the permanent adjustments are not perfect, the line of collimation will point up (or down) slightly when the instrument is levelled, and all staff readings contain a 'collimation error'. Since this error is proportional to the distance of the staff from the instrument, it will cause equal errors in sights of equal length. Consequently, since the carrying forward of the height depends on the difference between backsights and foresights, the errors will cancel out if, at each position of the instrument, the foresight and backsight distances are equal: i.e., I_1A should equal I_1B in Figure 6.1.

In practice, such errors will be kept small if the instrument is properly adjusted and the difference in distances is less than about 5 metres. Note, however, the errors introduced into intermediate sights do not cancel out, so these readings may have much larger errors unless the distance are short or the instrument is in very good adjustment.

The line of collimation can also deviate from the horizontal as a result of the tendency of light paths to bend in a vertical plane (see Chapter 10, Section 10.2). The effect of this varies with the square of the length of the line of sight and does not remain constant. With ordinary instruments, therefore, do not take sights longer than 50 metres.

When optical instruments are mounted at an awkward height, it is very easy accidentally to use one of the stadia lines (see Figure 4.1) instead of the main horizontal crosshair when taking a height reading. On a typical sight length of 30 m, this will introduce an error of approximately 15 cm, which should be apparent on checking. It is, of course, preferable not to make the mistake in the first place—so before taking a reading optically, always be sure that you can see all three horizontal hairs on the reticle.

On some instruments, the staff is seen upside-down in the telescope; do not correct this by holding the staff itself upside-down! Before starting work, study the scale carefully, both with the naked eye and through the telescope. Note the difficulty of distinguishing sixes from nines.

When levelling down a slope, it is easy to overestimate the length over which a reading can be made and to find that the instrument, when levelled, is either looking over the top of the staff or into the ground below it. For optical levels it is obviously necessary for the main crosshair (and perhaps the stadia hairs) to intersect the staff; and digital levels need to see a reasonable length of bar code above and below the crosshair, before they can take a reading.

To take a reading, the staff-holder faces the staff towards the instrument and stands behind it with one hand on each side so as not to hide the scale. It is clearly important that the staff should be held vertical, but this can be difficult to achieve, especially in windy conditions. Most staves (including all digital ones) have a cup bubble built into them, and the staff-holder must centre this and hold the staff steady* while the reading is being taken. When using an optical staff without a bubble, the observer can see if it is leaning sideways by means of the vertical hair in the telescope and can signal to the staff-holder accordingly. However, neither party can see if it is leaning forwards or backwards, so the staff-holder should therefore swing the top of the staff slowly towards and away from the instrument, passing through the vertical position. The observer then records the *smallest* numerical staff

---

* In windy conditions, it is helpful to have two poles about 1.5 metres long, and use these to 'brace' the staff by forming a tripod. If the staff is assembled from sections, it can also be helpful to remove any upper sections which are not needed for the reading.

reading that appears behind the crosshair, since this corresponds with the vertical position of the staff.

At change points, rest the foot of the staff on something firm and convex. If necessary drive a peg—preferably with a dome-headed nail driven into it—so that the foot of the staff is always in contact with the highest point when the staff is vertical. If no dome-headed nail is available, it is good practice to drive the peg in at a slight angle so that one corner is uppermost, rather than attempting to get the top face of the peg exactly level.

When levelling round a closed bay, it is important that the instrument should be moved (even if only slightly) for each set of observations. In Figure 6.2, for instance, it might not seem necessary for I_2 and I_3 to be separate physical positions. If they are not, though, there is the possibility that any misreading of the staff at point C will cancel out—the bay may then appear to close well, but there will be a height error at point C. This is even more important in a smaller bay such as C to E—the instrument *must* be moved from I_5 to I_6, even if this only involves pulling the tripod out of the ground and treading it back in again at almost (but not exactly) the same place.*

6.4 BOOKING

As mentioned above, most digital instruments can record their readings electronically, which makes booking unnecessary. This is undoubtedly efficient for surveyors who use a particular type of level frequently and are familiar with the relevant booking system; but for the first-time or casual user, it can be helpful to write the readings down in a book, even if they are also being recorded electronically. This is often useful when earlier readings need to be repeated, e.g., when a backsight has been taken and it then turns out to be impossible to read the required foresight—editing the electronically stored results can sometimes be challenging. It is therefore useful to know how to book levelling results manually—and, of course, there is no choice when an optical instrument is being used.

Figure 6.4 shows one standard method of booking. Successive rows of entries on the form refer to successive *staff* stations; the foresight from I_1 and the backsight from I_2 (both to staff position B, in Figure 6.2) are therefore booked on the same line. At each instrument station, the height of the line of collimation is obtained from the backsight, and then the reduced level of the next foresight is calculated by subtracting the relevant staff reading.

The booker, who may well also be the observer, books each reading in the field book as it is taken, using a pencil or ballpoint pen. Fill in a name for each staff station; the entries in the distance columns generally need only be

---

* To save time, the instrument can be left mounted on the tripod when this is done.

LEVELLING From: __Bench Mark A__ To: __Temporary BM C__ GROUP: __A2__   PAGE
Instrument: __Kern No. 4__ Observer: __A. Smith__   JOB: __Major__
Date: __28/3/03__   Booker: __B. Jones__   __Control__
Weather: __Light drizzle__ Checker: _____

Staff Position	Distance to Foresight	Foresight	Reduced Level	Distance to Backsight	Backsight	Line of Collimation
B.M. A	→	→	8.130	25	1.542	9.672
Point B	28	1.021	8.651	27	1.340	9.991
T.B.M. C	31	1.259	8.732	20	1.364	10.096
Point D	25	1.661	8.435	31	1.487	9.922
B.M. A	34	1.788	8.134			
	118	5.729	8.134	103	5.733	
			−8.130		−5.729	
			0.004	<Check>	0.004	
TBM C	>	>	8.730			
Point E						
TBM C						

Figure 6.4 Booking levelling results.

accurate to a metre or so, and can be judged by pacing.* Do not erase; make corrections by drawing a single line through the incorrect figures, leaving them legible, and writing the correct figures above them. A fair copy can be made later on another page if necessary—but take care to avoid copying errors, and do not destroy the original papers. Sign and date the work.

Immediately after booking a reading, verify it by again looking through the instrument; beware of gross errors of a metre or a tenth of a metre. Then, if the instrument has one, look at the sensitive bubble again to verify that the telescope is level.

When the bay is complete, the booker should add up the total of all the backsights (upwards movements) and of all the foresights (downwards movements), as shown in Figure 6.4. The difference between these two quantities should be the same as the calculated difference in height over the bay (in this case, 4 mm). If it is not, it means there is an arithmetic error somewhere on the booking sheet. This check is useful in that it might 'rescue' a bay which appears to have closed badly—conversely, it will also flag a bay which appears to have closed well as a result of two errors cancelling each other. Note, though, that it does *not* detect errors in the observations themselves. If the bay has closed acceptably and the arithmetic has been checked, the height of all points in the bay can be accepted. If desired, these

---

* The stadia hairs provide a useful alternative when pacing is impossible, e.g., if the line of sight passes over a stream.

heights can also be 'adjusted' to their most likely values, given any misclosure in the bay. In the case shown, point A has closed 4 mm higher than it should; the most likely assumption is that there has been a steady upwards 'drift' around the bay, meaning that the calculated heights for points B, C and D should be adjusted downwards by 1 mm, 2 mm and 3 mm respectively. The relevant adjustment is shown for point C in Figure 6.3, as the starting height for the next bay.

A further useful check is to add up the total distances for all the foresights and backsights in the bay, as shown in Figure 6.4. In this example the individual backsights and foresights from each instrument position (25/28, 27/31, 20/25, 31/34) are all individually within tolerance, but there has been a slight systematic bias towards having longer foresights than backsights, as shown by the totals. If the instrument has a collimation error, this accumulating difference will introduce errors into the recorded heights—and might, in this case, explain why the bay has not closed especially well.

6.5 PERMANENT ADJUSTMENTS

The permanent adjustments which can be made to a level ensure that the line of collimation is horizontal when the instrument has been levelled. Alterations to these adjustments should only be undertaken back at base, but it is sometimes useful to check them in the field, particularly if the instrument has just been subjected to a heavy impact.

A simple field test called the 'two-peg test' involves driving two pegs into the ground, a measured distance x (usually 25 m) apart, as shown in Figure 6.5. Set the instrument up approximately in line with (but not in between) the two pegs, and observe to the staff on each peg in turn (readings a and b in Figure 6.5). Move the instrument to the other end of the line (so that the peg which was near to the instrument is now the far one) and repeat the process, to obtain readings a' and b'.

The slope error of the instrument (in radians) is given by the expression

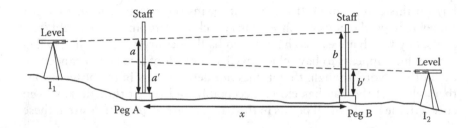

Figure 6.5 'Two-peg test' for a level.

$$\frac{(b - b') - (a - a')}{2x}$$

when all distances are expressed in metres and a positive value denotes an upwards slope. If the absolute value is less than 1×10^{-4}, then the instrument is fine; remember that even a value of 2×10^{-4} would cancel completely if the foresights and backsights are of equal length, and would only produce 1 mm of error if the sight lengths differed by 5 metres.

For information about tests and adjustments for a particular instrument, refer to the maker's handbook. Some digital instruments, for instance, have built-in software for computing the results of a two-peg test automatically.

6.6 PRECISE LEVELLING

A slower but much more accurate type of levelling is known as precise levelling, in which differences in heights are typically read to 0.01 mm, and bays might be expected to close to within 0.1 mm. The overall principles are identical to those described above, but some extra details are required to achieve the higher precision:

1. The staff is never rocked backwards and forwards as described above, but is made precisely vertical by means of a cup bubble. For a staff of more than 2 metres in length, this usually involves support by some form of tripod. The scale on the staff (optical or digital) is printed onto a strip of invar* held under constant tension in the frame of the staff, to minimise errors caused by thermal expansion.
2. For best efficiency (and certainly when using staves supported in tripods), a pair of staves is often used—one for the backsight, and one for the foresight. Since the two staves may not have identical offsets between the zero on the scale and the base of the staff, it is essential that bays always start and end using the same staff. This means that open bays must always have an even number of instrument positions. The smallest possible 'closed' bay (from a known benchmark to a new station and back to the known one) will therefore involve four instrument positions, so that the same staff is always placed on all the temporary benchmarks which are to be used as the starting point of a new bay. Thus, the bay from A to C and back in Figure 6.2 would be a valid bay for two-staff precise levelling, but the bay from C to E and back is too small.

---

* A ferrous alloy, typically consisting of 64% iron and 36% nickel, with a very low coefficient of thermal expansion.

The alternative approach is to use a single staff and carry it to and fro between the backsight and foresight positions—but this can involve a lot of walking, given the observation system described below. This approach is only viable if the staff is short enough to be held sufficiently steady by the observer—in practice this means the staff should not be more than 2 metres in length, which (in turn) means that the sight lines will be quite short when going up or down a steep hill.

3. A more elaborate system of observing is used to compensate for any changes in temperature which would cause even an invar staff to change slightly in length. The procedure is called 'BFFB' and involves reading a backsight, then a foresight, then a second foresight, and finally a second backsight. The time interval between readings 1 and 2 is kept small, and the interval between readings 3 and 4 is made to be the same length. Averaging the two results thus eliminates any temperature effects, assuming that the temperature is rising (or falling) at a constant rate. In an even more elaborate procedure, called 'alternating BFFB' or just 'aBFFB', the next instrument position observes foresight, backsight, backsight and foresight (i.e., FBBF), and this sequence of BFFB-FBBF is repeated throughout the bay.

4. Halfway through the observations at each instrument position, it is good practice to alter the foot-screws on the tribrach of a digital level slightly, so as to change the height of the instrument by 1 mm or so. This means that the two backsights and the two foresights are each taken to slightly different places on the staff, which helps to compensate for any nonlinearities in the digital interpolation system. Staves for optical precise levels usually have two scales, offset by one half of the range of the parallel plate micrometer, for the same reason.

5. The possibility of errors from any collimation error in the instrument is more important—either at one instrument position or cumulatively round a bay. To guard against this, the cumulative totals of distances to the backsights and foresights in a bay are recorded at each step and are steered to be as close to one another as possible (typically to within a few centimetres) throughout each bay. These distances are generally recorded by tachymetry (optical or digital) during the observation process, but steel tapes would also be used for planning the positions of the instrument and the staves.

6. The importance of finding a firm support on which to stand the staves is much greater. The staves are heavy and may settle by a millimetre or more while the readings are being taken, if they are resting on soft ground. Foot plates are generally used to support the staves when they can be placed on hard ground—they should be gently 'bedded in' to the surface, before the staff is placed on top. On softer ground, it may be possible to drive in a wooden peg, or a steel rod, having a domed top on which the staff can stand. Really soft ground (e.g., marshy

areas) should be avoided altogether. Whatever the surface, it is advisable to let each staff rest on its support for a minute or two before taking the first reading.

7. It is generally unwise to observe over a distance of more than about 30 metres, because of the atmospheric effects mentioned above. It is also unwise to observe to the limits (top or bottom) of the staves when going up or down a slope; if this is done on the outward half of a closed bay, it may prove impossible to complete the return half using the same number of instrument positions.

As with 'ordinary' levelling, the manual booking of precise levelling results is often a good idea, even if the level is capable of recording its readings electronically. Figure 6.6 shows a method for booking the results obtained from four instrument positions, running from a known benchmark to a temporary one. The readings are taken in the order suggested in

PRECISE DIGITAL LEVELLING OBSERVATIONS

From: Bench mark 'A'   To: TBM1   Date: 5/4/2013

Instrument: DNA03 No.1   Staves: Set 1   Weather: Overcast

Observer: A. Smith   Booker: B. Jones   Staffholder: C. Scott

St 1 → St 2		Backsight	Foresight	Distance (m)		Diff in elev (m)	
Ht 1	Temp	(m)	(m)	Back	Fore	Rise	Fall
BMA	pt1	•0.83976	1.90206	8.01	8.00		1.06230
	14Œ	0.83907	1.90139				1.06232
		1.67883	3.80345	8.01	8.00		1.06231
pt1	pt2	0.31566	•1.67778	7.72	7.68		1.36212
	13Œ	0.31585	1.67794				1.36209
		0.63151	3.35572	15.73	15.68		1.36210
pt2	pt3	•0.49522	1.97122	26.60	26.69		1.47600
	10Œ	0.49555	1.97159				1.47604
		0.99077	3.94281	42.33	42.37		1.47602
pt3	TBM1	0.80062	•0.49850	25.52	25.45	0.30212	
	11Œ	0.80066	0.49853			0.30213	
		1.60128	0.99703	67.85	67.82	0.30213	
	Totals	4.90239	12.09901		Totals	0.60425	7.80087
			Ð4.90239				Ð0.60425
Difference of Σ's			7.19662	Difference of Σ's			7.19662
Level of Bench mark A		257.362 m			x½		3.59831
Level of TBM1		253.76369 m		Difference of means			3.59830

Figure 6.6 Booking precise levels (digital instrument).

Point 3 above, starting with the one marked by the black dot in each case. If two staves are being used, the same staff would always be placed in the position suggested by the black dot, too.

When the four height readings and two distances have been recorded at the first instrument position (BMA to pt1), the difference between each backsight and its corresponding foresight is calculated and booked as a 'rise' or a 'fall', depending on whether the foresight station is higher or lower than the backsight station; in this case, it is clearly a 'fall' as the foresight readings are larger than the backsight readings.

On the final line for the first instrument position, the sums of the two foresights and backsights are written down (1.67883 and 3.80345 respectively), followed by the cumulative totals of the distances to the backsight and foresight (8.01 and 8.00) and the average of the rise or fall.

This process is repeated for three further instrument positions; note how the distances to backsight and foresight are being managed to ensure that the accumulating totals always stay within a few centimetres of each other.

When the observations have been completed, the totals of all eight backsight readings and all eight foresight readings are computed and recorded. The smaller number is then subtracted from the larger number, to leave a difference of 7.19662 metres, in this case. Likewise, the two 'rise' values and the six 'fall' values are also summed up, and a second difference is computed. If all the arithmetic has been done correctly, these two differences should be the same. As a final check (or as a way of determining which of the two differences is in error if they disagree), the three average falls are summed and the single average rise is subtracted from this total, to give (in this case) a net fall of 3.59830 metres which is booked as a 'Difference of means' in the bottom right corner of the form.

Finally, the exact change of height is computed by halving one of the 'Difference of Σ's' numbers to give (in this case) a net fall of 3.59831 metres. This may differ slightly from the 'Difference of means' result (and does, in this case) because of the rounding errors introduced when computing the averages. If the height of the first station is known, this can be recorded in the bottom left area of the chart, and the height of the final point can be computed using the computed change in heights. Note that the height of TBM1 has been recorded to a much higher precision than that of the start point; this is to ensure that rounding errors do not unnecessarily reduce the accuracy of the overall result, if a large number of bays are measured and combined.

Having completed a sheet such as that shown in Figure 6.6, the next task would probably be to level from TBM1 back to BMA, filling in a further sheet in the process. It is not necessary or desirable to use exactly the same change points for both tasks; but it can be helpful to leave pt2 in place on the outward journey, and to use it again on the way back, as shown in Figure 6.7. Then, if the two overall height differences do not tally, it will at least be possible to see whether the problem lies between pt2 and BMA or

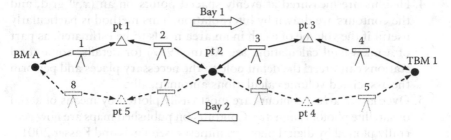

● Steel pin or stud in wooden peg
△ Staff support

Figure 6.7 Bays for precise levelling (plan view).

between pt2 and TBM1. Levelling the problem section for a third time will then indicate whether the error arose on the outward or return journey.

6.7 CONTOURS

Contours (Bannister, Raymond and Baker 1998) are the best method of showing variations in level on a plan; they can be thought of as the tide marks left by a flood as it falls by successive vertical intervals. Manual contouring is laborious; the following methods are used:

1. The contours are pegged out on the ground with a level and staff, and then surveyed (perhaps by a total station). The staff may have an adjustable marker set at the same height as that of the telescope above the required contour level, to speed its positioning.
2. A total station is set up and oriented as for mapping (see Chapter 3, Section 3.2), above a station of known height. A detail pole is set to the height of the instrument above its station, so that any height difference between instrument and reflector is always the same as the height difference between the station and the foot of the staff. The height difference from the station to the required contour is calculated, and the staff-holder follows the ground line which gives this height difference, with the instrument recording the staff position at suitable intervals. The contour is then plotted in the same way as other mapping detail.
3. The spot levels of points where the slope of the ground changes are taken, usually with a total station, and the contours are interpolated—usually by computer. This is a common method in engineering work; it is less accurate than method (2), but much quicker.

4. Heights are measured at evenly-spaced points on an (*x*,*y*) grid, and the contours are drawn by interpolation. This method is particularly useful if the volume of earth in an area needs to be estimated, as part of a mass-haul calculation; see Allan (1997) for details. Some total stations can 'steer' the detail pole to the necessary places and perform the associated volume calculations automatically.

5. Over larger areas, contours are most easily plotted by means of aerial or satellite photogrammetry. Contours on published maps are now generally plotted by digital photogrammetry (see Egels and Kasser 2001).

6. Contours may also be plotted by kinematic GNSS; see Chapter 7.

6.8 LEVELLING OVER LONGER DISTANCES

If the points are close to each other and only their relative heights are needed, these can be found as described above, using an arbitrary height datum (e.g., giving the first point a 'height' of 100.000 metres[*]). If, however, heights are required with respect to a national datum, e.g., for comparison with other points elsewhere in a country, then the scheme of levelling must include at least one point whose height is known with respect to that datum. This would typically be done by taking GNSS observations from one or more of the points (see Chapters 7 and 8). Although such observations are only accurate to a centimetre or so (meaning that the height difference may be in error by up to 2 cm), this is a much quicker (and possibly more accurate) approach than running a line of levels between two points which are several kilometres apart.

If no GNSS equipment is available, the scheme of observations can instead be tied into one or more nearby benchmarks of known height. The heights of benchmarks around Britain, for instance, are published by the Ordnance Survey (March 2013) at http://benchmarks.ordnancesurvey. co.uk—but these cannot now be fully relied upon as most of them are no longer checked regularly, as explained in Chapter 8. It is therefore good practice to use at least two (and possibly more) such benchmarks, to be certain that a reliable orthometric height has been established.

The most common benchmarks in the UK consist of a horizontal v-shaped slot cut into a brick or stone wall. They are used by inserting a bracket into the slot and standing a staff on the bracket. (The technical name for such a bracket is a 'bench', hence the term 'benchmark'.) In some places (e.g., on Ordnance Survey triangulation pillars) a brass 'flush bracket' is cemented into a wall to form a benchmark; this requires a more sophisticated bracket on which to stand the staff. Figure 6.8 shows how these two

---

[*] Avoid using a datum height of zero metres, as this may cause some other points to have negative heights.

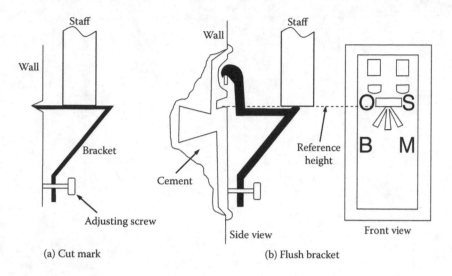

Figure 6.8 Use of Ordnance Survey bench marks.

types of benchmarks are used when levelling with a staff; the brackets typically incorporate an adjusting screw, so that (with the aid of a spirit level) the staff support can be adjusted to the same height as the reference mark, when the wall is not vertical.

Finally, trigonometric heighting (see Chapter 12) can be used to find height differences over longer distances. As discussed in Section 12.6, this technique is able to provide results which are accurate to about 4 mm per kilometre of separation, which makes it competitive with GNSS over distances up to about 5 kilometres.

Chapter 7

Satellite Surveying

7.1 INTRODUCTION

The human race has used objects in space to navigate for many years. In around 1000 BCE, the Phoenicians discovered that the maximum observable elevation attained by any particular star was a function of the observer's latitude, and they used this knowledge to navigate in an east–west direction across the Mediterranean Sea. With the advent of accurate chronometers in the 18th century, it became possible for navigators to determine their longitude as well as their latitude by means of astronomical observations.

The first use of satellites for position fixing began in 1964 with the Transit system, which consisted of six satellites in polar orbits and provided submetre accuracy after about two days of observations. The American GPS (global positioning system) came into service in the mid-1980s, and was originally set up as a military navigation aid; the Russian GLONASS (*global navigational satellite system*) started life in a similar way, a few years later.

At present, GPS and GLONASS satellites are both freely available to non-military users, and several other similar systems are in the process of being designed or commissioned. The European Galileo system now has four validation satellites in orbit, and is (at the time of writing) expected to be fully operational by 2020. The Chinese Beidou (or Compass) system is following a similar developmental path, and India is also planning a more limited system called IRNSS (Indian Regional Navigational Satellite System).

Collectively, these systems are now known as GNSS, or global navigational satellite systems. Most new navigational satellite receivers can receive signals from more than one of these systems, although GPS remains the single most important system for surveyors and other users. The technical details given in this chapter mainly therefore refer to GPS; the other systems mentioned above all work (or will work) in broadly similar ways.

The significance of GNSS to surveyors is profound. With care, it can now be used to measure positions on the earth's surface to subcentimetre accuracy after just a minute or two of observations; and it has the huge advantage (by comparison with earlier methods of surveying) that control stations do not need to be intervisible. This means that national networks of 'known' stations (provided around Great Britain by the Ordnance Survey and others) no longer need to be located on hilltops or tall buildings but can, for instance, be positioned on the verges of quiet roads.

7.2   HOW GPS WORKS

The GPS system consists of a set of about 24 satellites, each of which is in a near-circular orbit about the earth with a period of 12 hours* and therefore a radius of approximately 26,000 km. The orbits are all inclined at about 55° to the plane of the equator, and lie in six different planes, equally spaced around the equator. As a result, there are at least four satellites visible at all times everywhere on the surface of the earth, unless blocked by terrestrial obstructions. In most places and for most of the time, the number is greater than this, often up to eight or ten.

Each satellite broadcasts a set of orbital parameters (its 'ephemeris') which allow its position at any instant to be calculated to within about 20 metres, plus two digital signals whose 'bits' are transmitted at very precise times. By recording the time at which it receives the digital signal (and knowing the speed of light), a GPS receiver is able to determine how far it is away from the satellite (the 'pseudorange'†), and thus to position itself somewhere on a sphere with a known centre and radius.

When a second satellite is detected, another sphere is calculated; the locus of possible positions for the receiver becomes the circle of intersection between the two spheres. A third satellite provides yet another sphere, which will intersect this circle at just two points. One of these will typically lie many thousands of kilometres away from the surface of the earth—discarding this will give one possible position for the receiver.

At this point, the principal error in the calculation is caused by the clock in the receiver (the satellites have atomic clocks, which are highly accurate). Because light travels at 300 mm/s, an error of just 1 microsecond in the receiver's clock will cause an error of 300 metres in the calculated radii of all the spheres, and thus a large error in the calculated position. For this reason a fourth satellite must be detected, and a fourth sphere calculated— the radii of all four spheres are then adjusted by an equal amount, such that they all touch at one single point. This point is taken as the position of

---

* Strictly, 12 *sidereal* hours (a sidereal day being the time for the earth to complete one revolution with respect to the stars, rather than the sun). A given constellation of satellites therefore recurs twice each day, and about 4 minutes earlier on each subsequent day.

† The pseudorange of the satellite is the distance calculated by measuring the time when the digital signal is received, not allowing for clock errors in the receiver or satellite, or for delays caused by the earth's atmosphere.

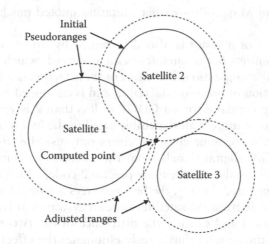

Figure 7.1 Receiver clock adjustment in 2-D satellite surveying.

the receiver, and the required adjustment in the radii (divided by the speed of light) is taken to be the receiver clock error. The receiver's clock is then adjusted accordingly, and this error is eliminated from the pseudoranges.

If more than four satellites are visible, the extra information can be used to provide redundancy in the calculation, and the receiver will report a position based on the best fit of the available data.

If only three satellites are visible, some systems will also provide a 'two-dimensional' (2-D) solution, by assuming that the receiver is at sea level. Figure 7.1 is a plan view of the earth's surface which shows how three satel-lites can provide such a 2-D solution, and correct the receiver clock error. The three solid circles are the loci of all points on the earth's surface which are the appropriate distance from each satellite, as calculated from the pseudoranges. As can be seen, there is no one place on the earth's surface which lies on all three circles. However, if the receiver's clock is running slow, this would cause it to underestimate its distance from each satellite; advancing the receiver's clock appropriately and recalculating the ranges gives the three dotted circles, which do meet at a point.

The method described above will enable a single GPS receiver to calcu-late its so-called 'navigational' position to within about 10 metres.* This accuracy can be improved to better than 1 metre by leaving the receiver in the same place for an hour or more, and averaging the readings. Note that these figures have improved significantly since the US military withdrew 'selective availability' (the deliberate downgrading of the data provided by

---

* The accuracy in vertical position is always about 2.5 times poorer that the accuracy in horizontal position, due to the fact that the satellites being observed are always above the plane of the horizon. This factor also applies to all subsequent accuracies described in this chapter.

the satellites) in May 2000—earlier literature quoted much higher errors than this.

The accuracy of a result is also determined by other factors, such as the relative positions of the satellites being observed, which will affect the geometry of the computation. The effect of these factors is referred to as the geometric dilution of precision (GDOP),* and is expressed as a multiplying factor for the potential error—a GDOP of less than 2 is very good, but it could rise to 20 or more if all the visible satellites lie in a near-straight line across the sky. Also, some cheap receivers only use the so-called 'coarse acquisition' (C/A) digital signals from the satellites to compute pseudoranges, while others also use the more precise P-code, which has a 10-times higher 'chipping rate.'† Finally, the best receivers are dual frequency—they receive the P-code from the satellites on carrier waves at two slightly different frequencies‡ and can use the difference of the two resulting pseudoranges to estimate (and thus largely eliminate) the effect of the earth's atmosphere on the speed of propagation of the signals.

The accuracy discussed above refers to the *absolute* position of the receiver on the surface of the earth and is considerably higher than anything which could be achieved prior to the 1960s, using astronomical observations. However, it is insufficient for many engineering purposes, which typically require the *differences* between stations to be known to a few millimetres. For this reason, surveyors tend to use differential GNSS, or DGNSS, which is described in the next section.

7.3 DIFFERENTIAL GNSS (DGNSS)

The factors which most affect the accuracy of a single high-quality GNSS receiver are errors in the positions of the satellites, errors in the satellite clocks, and the effects of the earth's atmosphere on the speed at which the satellite signals travel. If two such receivers are within, say, 10 km of each other, the effects of these factors will be virtually identical and the *difference vector* in their navigated positions will be correct to within a decimetre or two. If the distance between the receivers is greater than this, the accuracy of a simple difference calculation is degraded by the fact that the two receivers will be observing the same satellites but from somewhat different angles—so that the errors mentioned above will all have slightly

---

* Sometimes broken down into the 'positional dilution of precision' (PDOP) for the accuracy of the positioning on a horizontal plane, and 'vertical dilution of precision' (VDOP) for height information, due to the general difference in accuracy of these two calculations, mentioned above.
† The rate at which bits are transmitted in the satellite's binary signal.
‡ These are known as the L1 frequency (1575.42 MHz) and the L2 frequency (1227.60 MHz).

different effects on the calculated positions of the two receivers. The satellite clock and position errors can be eliminated if one of the receivers (known as the 'base station') is at a known position for an extended period of time—the processing software corrects the positions of the satellites using the data recorded by that receiver, and then applies those corrections to the other receiver (known as the 'rover'). This can be done in real time if the base station is equipped with a radio transmitter with which to send its corrections to the rover; alternatively, both stations can simply record their observations, for subsequent post-processing.

The fact that signals to the two receivers are passing through different parts of the earth's atmosphere, and will therefore suffer different propagation effects, is harder to correct*. This effect ultimately constrains the overall accuracy of differential GNSS (DGNSS) to about 2 mm in the horizontal plane and 5 mm vertically for every kilometre of separation between the two receivers (i.e., 2 and 5 parts per million), up to the point where the two receivers can no longer see the same satellites.

The final precision of differential GNSS is achieved by measuring the phase of the carrier wave onto which the P-code is modulated. The chipping rate of the P-code is 10.23 MHz, which means the bits in the signal are about 30 m apart. By contrast, the L1 carrier wave has a frequency of 1575.42 MHz, and thus a wavelength of about 19 cm. Interpolation of the phase of the carrier signal will yield a differential positional accuracy of a few millimetres, provided it has been possible to use the P-code to obtain a result to within about 20 cm beforehand. If not, the carrier phase information cannot be used because of the uncertain number of whole wavelengths between the satellite and the receiver. The attempt to determine the number of whole carrier wavelengths is called 'ambiguity resolution'. It is usually possible to resolve ambiguities when the receivers are up to 30 km apart, given a good GDOP and enough observation time—and it is usually unwise to attempt it† if the receivers are more than 50 km apart, because of the unknown differences in atmospheric delays along the two paths. Note, therefore, that the term 'DGNSS' can imply a wide range of relative positioning accuracy, from about 2 mm up to 2 decimetres or so.

A final factor which is important at the top level of precision, is 'multipath', i.e., the reception of signals which have not come directly from the satellite but which have bounced off (for instance) a nearby building—this can cause errors of up to half a metre in the calculated position of the

---

* Some commercial solutions to this have now been developed and are discussed below.

† Processing the data under these circumstances may yield a seemingly plausible solution—which might, in fact, be incorrect by one or more whole wavelengths. Software from responsible suppliers will warn a surveyor against using results unless the statistical likelihood of their correctness is high. Even then, however, it is impossible to guarantee that the calculation has yielded the correct result.

receiver. For this reason, surveying GNSS stations should always be sited well away from buildings and large metal objects. In particular, DGNSS cannot always be relied upon to produce accurate results in the middle of a construction site; it is often good practice to use DGNSS to fix control stations around the edge of the site, and then to use the more conventional surveying methods within the site.

7.3.1 Base Stations for Differential GNSS

One requirement for the accurate computation of a DGNSS difference vector is that the absolute position of the base station is known to within about 1 metre before the calculation is done. If a completely 'local' co-ordinate system is to be used for a project, it is perfectly acceptable to base the whole system on a point which has been fixed as a navigational solution, provided it is observed for long enough to fix it to that accuracy. All difference vectors built out from that point will be of high accuracy, and all points fixed using those vectors will also therefore have an absolute accuracy of less than 1 metre, so can in turn be used as base stations for further vectors.

Often, however, it is necessary to tie in new GNSS stations to a country's national mapping system. This can be done in four different ways, using three different types of 'known' stations.

7.3.1.1 Passive Stations

Many countries, including the UK, provide a network of stations with known (and published) co-ordinates. These are often sited on roadsides or other public places, and so can be occupied without obtaining permission. Using one or (preferably) more of these stations as base stations will tie all new stations into the national co-ordinate system to a reasonable level of accuracy.

7.3.1.2 Active Stations

In addition to passive stations, several organizations maintain 'active' base stations, at known positions. These record GNSS data which are subsequently published (usually via the Internet) and which can be downloaded for post-processing in conjunction with data recorded by a roving receiver. This system allows users with only one GNSS receiver to carry out DGNSS, and increases the productivity of users with more than one receiver. The format of the data is normally RINEX (receiver-independent exchange format), which is the standard for transferring GNSS observations between different manufacturers' equipment. The published co-ordinates of these active stations are of a higher general quality than those of the passive stations described above, so the results will be correspondingly more accurate.

Before using this service, it is wise to check the frequency at which the chosen active station records its observations (typically once every 15 seconds), and to set your own receiver to record at the same frequency; this simplifies, and improves the quality of, the subsequent post-processing. Be prepared also to return from recording your own observations only to find that they cannot be used to full effect because the nearest active station wasn't working that day!

The fact that the base and roving stations may be using different types of antennae may also cause problems, as they will have different offsets. The documentation for the post-processing software should explain how to allow for this—but any error in inputting this information will potentially go undetected. As a check, download some further data from another active station, with yet another antenna type, and check that the two differential vectors produce compatible results.

It is in any case good practice to download data from several (five, say) nearby active stations, and treat them all as 'base stations' when post-processing the data. As well as detecting systematic errors of the type mentioned above, this will help cancel out atmospheric effects—and the accuracy to which the differential vectors meet together at a single point will give a good estimate of the accuracy of the observations. Ideally, these active stations should form a 'ring' around the point being measured—this is made easier in the UK by the fact that at least half of the 100-odd active stations maintained by the Ordnance Survey are sited on, or near, the coast.

7.3.1.3 Broadcasting Stations

An emerging service in several countries is the permanent installation of GNSS receivers which act as base stations, and broadcast their data via short-wave radio to any nearby receiver. Surveyors who have paid to use the service, and who have suitably equipped receivers, can use this information to show their position to within a centimetre or so in real time. This system is typically used at airports, enabling DGNSS to be used as a precision landing aid.

7.3.2 Network Real-Time Kinematic Services

More recently, commercial services have started to appear which maintain a network of receivers at known points, and use the data from these receivers to estimate the current satellite errors and also to generate a model of how the atmosphere is delaying signals from those satellites. When a suitably equipped roving GNSS receiver starts to operate, it first determines its 'navigational' position, which it reports (via a mobile phone) to the service. The service then uses its data to compute the errors (satellite and atmospheric) which a fixed base station would experience if it were placed at the position reported by the roving receiver, and gives the receiver the appropriate corrections; effectively, it creates a 'virtual active station' within a few metres of the receiver. This allows very quick position fixing by a single receiver, to an accuracy which is mainly governed by the quality of the atmospheric model.

7.4 USING DGNSS IN THE FIELD

Differential GNSS relies on the same satellites being observed at the same time by the two receivers. If one receiver is recording while the other is not, those observations will be unusable. There are a number of ways of using DGNSS in practice, depending on the size and purpose of the survey. The principal ones are as follows.

7.4.1 Static

When the two receivers are more than about 15 kilometres apart, it is necessary for them to remain simultaneously in position for an hour or more, recording observations every 15 seconds or so. The time period allows the satellites to move through significant distances and for a larger number of satellites to be observed by both receivers simultaneously—and the number of observations ensures a good chance of resolving ambiguities if that is possible, or of obtaining a well-averaged result if it is not. Static survey is usually used for the establishment of new control stations in an area well away from any existing 'known' stations.

7.4.2 Rapid Static

If the distance between the receivers (the 'baseline') is less than about 15 km, the observing time can be reduced because the atmospheric effects will be nearly identical for each receiver. The time required depends on the length of the baseline, the number of satellites, the GDOP, and the algorithms in the receivers. The instruction manual should give advice on the observation time required; failing that, about 10 minutes is probably prudent in most cases. With lines shorter than 5 km, five or more satellites, and a GDOP of less than 8, five minutes will probably be adequate.

7.4.3 Stop and Go

In this procedure the roving receiver makes a rapid static fix at its first station, and is then moved to other stations while maintaining a lock on the satellites which it is observing. Subsequent points can then be fixed very quickly, in about 10 seconds. This procedure is suitable for collecting the positions of a large number of points in open country—but if fewer than four satellites can be tracked at any point, a new 'chain' must be started by doing another rapid static fix.

7.4.4 Kinematic

This procedure also starts with a rapid static fix,* after which the roving receiver moves continuously, recording its position at regular time intervals (perhaps as frequently as once per second). As with 'stop and go', satellite lock must be maintained at all times. This technique is typically used for surveying boundaries and other line features.

7.4.5 Real-Time Kinematic (RTK)

If two suitably equipped receivers are less than about 5 km apart and have a near line of sight between them, it is possible for the base station to transmit its position and observations to the roving receiver using a short-wave radio. The roving receiver can then carry out the DGNSS calculations in real time, and display its current position in the WGS84 or ETRS89 co-ordinate system (see Section 7.6). If a suitable transform and/or projection has also been downloaded, the roving receiver can also display its position in the local co-ordinate system. Using RTK, the operator of the roving station can be confident that enough observations have been recorded to resolve ambiguities while still out in the field, so can carry out rapid static or stop-and-go procedures more quickly. In addition, RTK can be used to set out a station at a predetermined location, albeit without any redundancy or independent check of its accuracy, as discussed in the next section.

Whichever method is used, there are some fundamental rules which should be followed when using GNSS to maximise the chances of accurate results:

1. Avoid using satellites which are at a low elevation (less than 15° above the horizon, say), as the signals from these satellites will be greatly affected by atmospheric effects, due to their long path through the atmosphere. Most surveying receivers will ignore all such satellites, by default.
2. Avoid working close to large buildings. In the northern hemisphere, a building to the south will tend to block out some visible satellites, while buildings anywhere else may cause multipath effects. The combination of these two effects is potentially serious!

---

* Increasingly, the processing software is capable of resolving phase ambiguities even while the roving receiver is moving—but a user would normally wish to wait at the first point until enough readings had been taken to resolve ambiguities, in order to fix the position of that point.

3. Working beneath the canopy of a tree can also block the signals from satellites. When working in stop-and-go or kinematic mode, just passing briefly beneath the canopy of a tree can cause loss of lock, resulting in reduced accuracy for all subsequent readings in the chain until the phase ambiguities are resolved again.

4. Some GNSS processing systems allow a surveyor to check in advance what the GDOP of the satellite constellation will be at the time when it is planned to take readings. If only one satellite constellation is available, this simple precaution can avoid long periods wasted out in the field waiting for the GDOP to improve to the point where useful readings can be taken. For receivers capable of using more than one system, the total number of satellites makes such delays unlikely.

7.4.6 Building a Network of Stations

Usually, the goal of a satellite survey is to establish the precise location of a number of fixed stations in the field. Logically, this is achieved by fixing the position of an 'unknown' station with respect to one or more 'known' ones using DGNSS. The unknown point then becomes known and can be used (if required) as the basepoint for further differential vectors, to find the position of other unknown points.

In practice, there is no need for the sequence of observations to follow this logical order—it is simply necessary for the results to be *processed* in that order. Nor is it necessary for one receiver always to act as the base station, while the other always acts as the roving station; it is quite permissible for them to 'leapfrog' each other, taking the role of base and rover respectively as a chain of points is visited.

It is, however, important to plan in advance what readings need to be taken, and then to plan a sequence of movements for the receiver(s) to ensure that they are, in fact, taken. A clear written record of what observations have been taken by each receiver (together with a note of the height of the antenna above the station) will also greatly simplify the subsequent processing and archiving of the data. A form for this purpose is given in Appendix G.

7.5 REDUNDANCY

Attentive readers of this book will be aware of the need for redundancy in all surveying measurements. Although properly post-processed DGNSS results are the average of many individual 'observations', there is still the possibility of a systematic error (e.g., the height of one receiver not being correctly recorded) which will cause an erroneous result.

The straightforward solution to this is to establish each new station by setting up DGNSS vectors from at least two 'known' stations. A gross error will then quickly be detected if the two vectors do not meet at almost the same point. Furthermore, the post-processing software supplied with the GNSS equipment will probably contain a least-squares adjustment facility, which will find a 'best' position for each unknown point, based on the co-ordinates of the known points and the DGNSS vectors which have been collected. Clearly, collecting more 'redundant' vectors will result in a more accurate result with a smaller chance of an undetected gross error.

If time is limited, redundancy can be achieved by conducting a GNSS 'traverse', similar to the conventional traverse described in Chapter 2. The first unknown point is fixed with respect to an initial known station, and is then used to fix the next point. This then becomes the base station for the third point, etc., finally finishing on another known point (preferably not the one where the traverse started). If the position of the final point as calculated by the traverse is in good agreement with its known position, it can be assumed that all has gone well. This approach does, however, have two drawbacks:

1. It is possible that a satisfactory result masks two errors which have cancelled out, e.g., the height of an antenna was wrongly measured at one station along the traverse.
2. If an error is detected, it will not be possible to determine the leg in which it occurred, and another visit to the site will be necessary. It may, therefore, be sensible to take more redundant readings on the first visit, since this might allow a faulty reading to be eliminated without another site visit.

Ultimately, though, all surveyors should be aware that GNSS has the tendency to be a 'black box' science, in which a large amount of information is collected, and may not be fully checked by the user. There is a distinct possibility of some overall systematic error in the collection or processing of GNSS information (the inappropriate use of a program, the wrong settings in a transformation, or even a software bug) which might cause a completely undetected error in the final result. If absolute confidence is required in a set of new stations for a major project, it is strongly advised that a few checks are made by conventional surveying techniques. The remeasurement of some distances using EDM, for example, is perhaps less accurate than DGNSS—but will clearly show if a scaling error has inadvertently been introduced into the GNSS results. This is simply the modern equivalent of the old practice of pacing a distance which has been measured by tape, to check that the number of complete tape lengths has not been miscounted.

Height information obtained from DGNSS needs particular care—partly because it is less accurate anyway, and partly because of the nature of the co-ordinate system used by GNSS, which is described in the next section. If the heights of stations fixed by GNSS are to be relied upon, it is strongly recommended that the relative heights of some stations in the network are checked by conventional means, such as levelling (Chapter 6) or by trigonometric heighting (Chapter 12). Again, these conventional methods may be less accurate than GNSS—but if any discrepancies are too large to be accounted for by their inaccuracy, then there is clearly a potential problem with the GNSS results.

7.6 PROCESSING GNSS RESULTS

The ephemeris information of all GNSS satellites, and thus the navigational position of a single GNSS receiver, is expressed in terms of a co-ordinate system called WGS84.[*] In its basic form, WGS84 is defined as a set of right-handed orthogonal axes, with its origin at the centre of mass of the earth, the x- and y-axes lying in the equatorial plane, and the positive z-axis passing through the north pole.

The exact orientation of the axes was originally set up to coincide with the equatorial plane as defined by the Bureau Internationale de l'Heure at the very start of 1984, with the x-axis passing through the 'prime meridian' (approximately the Greenwich meridian) at the same instant. Subsequently, the orientation of the x- and y-axes has been defined such that the mean drift of all the tectonic plates on the earth's surface is zero with respect to them.

This Cartesian system is complemented with a biaxial ellipsoid[†] of defined shape (close to the overall shape of the earth), with its centroid at the origin and its axis of rotational symmetry lying along the z-axis, as shown in Figure 7.2. This gives a more natural way of defining a point on the earth's surface, in terms of its geodetic latitude (the angle ϕ), its longitude (the angle λ), and its height above the ellipsoid (h). These two co-ordinate systems are called Cartesian and geodetic respectively, and are explained in more detail in Chapter 8.

This is a properly 'global' co-ordinate set for a global positioning system—but unfortunately it does not suit any single country particularly well, for two principal reasons:

1. The positions of 'fixed' points on the earth's surface (e.g., concrete blocks set into the ground) do not have constant co-ordinates in the WGS84 system, because of continental drift. On the Eurasian tectonic plate (which includes the UK), this drift is in excess of 2 cm per year—in some parts of the world, it is up to 10 cm per year.

---

[*] World Geodetic System, 1984.
[†] The three-dimensional shape created by spinning an ellipse about its minor axis.

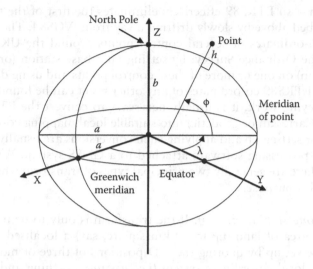

Figure 7.2 Cartesian and geodetic co-ordinate systems in WGS84.

2. The heights reported by GNSS measurements are heights above the surface of the WGS84 ellipsoid, which is in general not coincident with, or even closely parallel to, the surface of zero height (the 'geoid'—see Chapter 8) in any particular country. Thus, the difference in ellipsoidal heights between two stations measured by differential GPS might be somewhat different from the difference in their orthometric heights, as measured using a level. A naïve surveyor might be surprised to find that water can flow from a point with a low WGS84 height to another point with a greater WGS84 height!

To make proper use of GNSS results, it is therefore necessary to understand about the relationship between WGS84 and the heighting and mapping co-ordinate systems used within a particular country.

In Europe, the first step was to select a number of ground stations around Europe (all on the Eurasian tectonic plate) and to establish a co-ordinate system similar to WGS84 which is defined to be the best fit between those physical stations and their accepted WGS84 co-ordinate values at the start of 1989. This definition forms part of a system called ETRS89,[*] and the set of ground stations which 'realise' it (i.e., make it real, and available to surveyors) is called the European Terrestrial Realisation Frame, or ETRF. No geologically stable ground station in Europe moves within this co-ordi-

---

[*] European Terrestrial Realisation System, 1989. This system also specifies that the WGS84 ellipsoid should be used to convert between Cartesian and geodetic co-ordinates.

nate system*—so ETRS89 effectively eliminates the first of the two problems described above by slowly drifting away from WGS84. The definitive ETRS89 co-ordinates of several control points around the UK are published by the Ordnance Survey; by setting up a base station (or using an active station) on one or more of these control points and using differential GNSS, the ETRS89 co-ordinates of any other point can be found.†

For surveying work, it is usually necessary to convert the ETRS89 co-ordinates (Cartesian or geodetic) into suitable local mapping co-ordinates. Facilities for setting up and applying such conversions are usually provided in the post-processing software attached to a GNSS system. Within such software, there are generally two principal types of transform which can be set up for this purpose:

1. *The 'one-step' transform.* If the transform is only to be used over a small area of land (up to 10 km square, say) a localised transform can be set up by quoting the 3-D positions of three or more‡ points in the local co-ordinate system (i.e., easting, northing and height in the country's mapping system, or a site co-ordinate system) and also in ETRS89. Provided the points are well distributed over the area in which the transformation is to be used,§ the errors inherent in this type of transform are small by comparison with the errors inherent in differential GNSS. All other points whose positions have been found in ETRS89 can then be processed through the transform to find their local co-ordinates.

2. *The 'classical' transform.* In all mapping projection systems there is a scale factor, which varies from place to place, by which a distance measured on the ellipsoid must be multiplied before it can be plotted on the projection (see Chapter 9). The 'one step' transform can accommodate this, but assumes that the scale factor is constant over the area for which the calculated transformation is to be used. If the area is too big for this assumption to be valid, it is necessary first to use the so-called 'classical' or Helmert transform, which transforms the WGS84/ETRS89 Cartesian co-ordinates into another set of Cartesian co-ordinates, centred on whatever ellipsoid has been used for the mapping projection (often Airy, in the UK). These are

---

* The published co-ordinates of a station may change from time to time, though, as a result of more accurate measurements of its position. This is discussed later in this chapter.

† Although the ETRS89 co-ordinates of the base station will differ by a few centimetres from its WGS84 co-ordinates, this difference is too small to impair the accuracy of the DGNSS calculation described in Section 7.3.1.

‡ If more than three points are available, a least-squares fit will be used to generate the best transformation between the two co-ordinate systems.

§ The accuracy of the transform will be compromised if the points lie in a near-straight line. This will not matter, however, if the area of the survey is itself a near-straight line, e.g., a pipeline or road.

then converted to geodetic co-ordinates (latitude and longitude) for the relevant ellipsoid (see Chapter 8). Next, the appropriate projection system (and, sometimes, a co-ordinate shift system) is applied to give map co-ordinates of easting, northing and height above the local ellipsoid, as explained in Chapter 9. Finally, a 'geoid model' is needed to convert the height above the local ellipsoid into a height above the geoid (called an orthometric height). A complete roadmap of this process is shown in Figure 7.3.

A classical transform will produce valid results over a much larger area than a one-step transform. In order to set it up, it is necessary to know the co-ordinates of three or more points in both systems—i.e., in WGS84 (or a local variant of WGS84 such as ETRS89), and also in the local mapping projection system. Typically, a surveyor in Europe will have measured the ETRS89 co-ordinates of three or more stations, and will also know their map (or grid) co-ordinates, plus their ellipsoidal heights. The transform software can use this data, plus details of the ellipsoid and projection method used to make the map, to generate two sets of Cartesian co-ordinates for the stations, and then use a least-squares process to generate the 'best fit' Helmert transform between them.

Both transforms allow for full 3-D rotation and translation, to convert from one co-ordinate system to the other. In addition, they both make provision for a scale factor to be introduced, to give the best possible fit between the two systems. This can be useful, but the scale factor must then be applied to any length which is to be converted from a distance on the ground to one

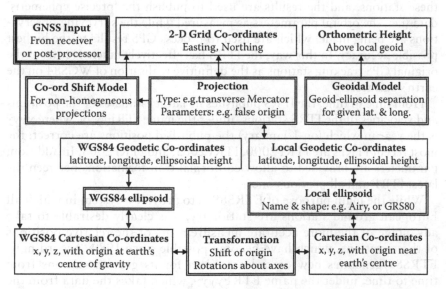

Figure 7.3 Converting between WGS84 and local co-ordinate systems.

in the local co-ordinate system, or vice versa. In the case of the classical transform, this scale factor is *additional* to any scale factor implied by the projection method—so when setting up classical transforms, it is often better to insist that the transformational scale factor is kept at unity.

It is clear from the descriptions above that the 'one-step' transform is the easier one to use, provided the area of application is sufficiently small. Exact details of how to set up and apply both these types of transforms will be found in the user manual for the post-processing software provided by the GNSS supplier. The mathematical details of conformal transforms are explained in Chapter 8, Section 8.5.

7.7 THE INTERNATIONAL TERRESTRIAL REFERENCE SYSTEM

The orbital parameters of navigational satellites are established by observing them from ground stations whose WGS84 co-ordinates are precisely known. In the early days of GPS, this was done exclusively via the GPS 'control segment', whose six tracking stations thus effectively defined the realisation of WGS84 on the earth's surface. Subsequently, the International Earth Rotation Service (IERS) established the International Terrestrial Reference System (ITRS) in terms almost identical to those of WGS84, and now measures the positions (and relative velocities) of a much larger network of ground stations with respect to this co-ordinate system with ever-increasing accuracy. All the GPS satellites are now tracked by a number of these stations, and the results are used to publish the 'precise ephemeris' data (i.e., the orbital parameters as measured while the surveying observations are being made) which are used to process GPS results to the highest possible accuracy. In this way, the ITRS has effectively taken over from the original GPS tracking stations as the definitive realisation of WGS84 on the earth's surface.

From time to time, the IERS publishes its latest estimates of the positions and velocities of the ITRS stations, under the name ITRF*yyyy*, where *yyyy* is the year in which (on 1 January) the published positions are correct: the most recent ITRFs are ITRF2000, ITRF2005, and ITRF2008. In addition, parameters are provided to allow conformal transformations between the latest ITRF and all previous ones.

While the main purpose of ETRS89 is to provide a system in which all European ground stations are stationary, it is clearly desirable to take advantage of the more accurate measurements of the relative positions of these stations, which have become possible since 1989. Accordingly, ETRS89 generates new co-ordinate data for its ground stations from time to time, under the name ETRF*yyyy*, which takes the data from the corresponding ITRF*yyyy* and then 'winds the clock back' to find the best estimate of where the stations must have been in 1989. The current ETRF is ETRF2000, which is (of course) based on the data published in ITRF2000.

7.8 FURTHER DETAILS OF GPS AND GALILEO

The GPS system is still the mainstay of satellite navigation around the world, so some further details of how it works will be of interest to surveyors; the other systems work in similar ways, but with slightly different details.

The GPS system consists of three so-called segments:

1. *The Control Segment.* This comprises the computing power neces- sary to track the satellites, and to predict their orbits for 24 hours in advance using a highly sophisticated model of the earth's gravi- tational field. Tracking is done by a network of six tracking stations around the world, which then feed information back to the main con- trol centre in Colorado. The orbital predictions are uploaded to the satellites every 24 hours. Corrections to the satellites' clocks are also uploaded by the control segment. It is interesting to note that the four atomic clocks carried by each satellite run at a slightly different rate once the satellites have been launched, due to the speed at which the satellites are travelling (about 4 kilometres per second) and the effects of relativity.

   If a satellite's orbital parameters drift outside certain limits, the satel- lite is marked as 'unhealthy', so that GPS receivers will not use it; the satellite's orbit is then adjusted using small on-board rocket motors, and its new orbit is tracked for a while. Once the new orbital param- eters have been uploaded, the satellite is returned to a 'healthy' state.
2. *The Space Segment.* This is the set of (nominally) 24 satellites, each of which receives and stores its predicted orbital parameters, and trans- mits this and other information to . . .
3. *The User Segment.* The rest of the GPS community.

Despite the huge expense of deploying and maintaining the satellite network, an expense borne by the US Department of Defense, the US Government is currently committed to maintaining a level of free civilian access to the system.

7.8.1 The GPS Signal

Each of the satellites transmits signals on two carrier frequencies, both derived from a fundamental oscillator running at 10.23 MHz. The two frequencies are called L1, 1575.42 MHz or 154 times the fundamental, and L2, 1227.60 MHz or 120 times the fundamental. As explained earlier, the purpose of having two frequencies is to enable the correction of errors caused by ionospheric effects.

Each satellite also transmits a number of digital signals modulated on the L1 and L2 signals. It is important to realise that, since the signal strength is so weak in relation to the background noise and all the satellites transmit at the same frequencies, a particular transmission can only be recognised by knowing in advance what code is modulated onto it, and thus what to 'listen' for.* These codes are:

1. The coarse acquisition or C/A code, a pseudo-random† bit sequence of length 1023 bits, different for each satellite and with a repetition time of 1 ms. The C/A code enables the receiver to distinguish between transmissions from the different satellites. It is used in low-cost navigation receivers as the basis for measurements. The bits are released at a rate of 1.023 Mb/s (the so-called chipping rate, derived from the fundamental oscillator), so one bit corresponds to a distance of approximately 300 m. The C/A code hence gives access to what is known as the standard positioning service.

2. The precise or P-code, also a pseudo-random bit sequence, but at ten times the frequency of the C/A code, with a chipping rate of 10.23 Mb/s. The cycle length of the complete P-code is in excess of 37 weeks, and each of the satellites is allocated a different 'week', so that each satellite effectively has its own P-code. The code for all satellites is reset every week at midnight on Saturday/Sunday. Contrary to what is said in much of the early literature, the P-code is *not* restricted to military use in itself—the generation algorithm is, and always was, in the public domain. However, its use can sometimes be denied to the civilian user by the substitution of an encrypted form of P-code known as the Y-code. The encryption algorithm is not publicly available. Encryption is known as anti-spoofing or AS, since its purpose is to prevent an enemy force from setting up a 'spoof' transmitter which could make the US military receivers indicate false positions.

3. The navigation message, a digital data stream running at 50 b/s. This message contains, amongst other things, orbital information (called the ephemeris) for the transmitting satellite, repeated every 30 s, and less precise 'almanac' information to tell the receiver which other satellites are likely to be visible. Reception of the almanac for the whole constellation takes 12.5 minutes.

---

* Somewhat similar to hearing one's name spoken on the other side of a room full of people talking.
† The bit pattern in the code looks as though it is random, and would pass most 'randomness' tests—but it is in fact entirely predetermined.

The navigation message and the P- or Y-code are carried on both L1 and L2 frequencies, whereas normally the C/A code only appears on L1. This makes acquisition of L2 signals difficult when AS is present, but the manufacturers of survey receivers have developed ingenious ways of avoiding the problem.

7.8.2 Ionospheric Effects in GPS

As mentioned earlier, these can be estimated by comparison of pseudoranges measured on both L1 and L2 frequencies. (It should be noted that the ionosphere *delays* the code parts of the signals, but *advances* the carrier phase by an equal amount.) Single-frequency receivers thus have greater difficulty in resolving phase ambiguities, since they have no estimate of the ionospheric effect, which dual-frequency receivers can derive from the different delays in the code transmissions on the two frequencies.

7.8.3 GPS Time

The fundamental measure of time used in the world is called universal co-ordinated time, or UTC. The length of a UTC second is defined by the decay rate of caesium. To ensure that midnight continues to occur in the middle of the night on average, 'leap seconds' are introduced into UTC when necessary, which means that the final minute in June or December occasionally lasts for 61 seconds instead of 60.*

GPS time is measured in weeks and seconds from 0:00:00, on Sunday, 6 January 1980. It is established by averaging the clock readings from all the satellites plus a ground-based master clock, and is then steered so that its seconds increment within one microsecond of UTC seconds. However, GPS time has no leap seconds, so now runs ahead of UTC by more than 15 seconds. A further complication is that the 'week counter' is only 10 bits long, so 'rolls over' to zero every 1024 weeks. This occurred for the first time on 22 August 1999, causing problems in many receivers. It will happen again on 7 April 2019.

A simple method for finding the GPS week number (useful for downloading precise ephemeris data, as described in Section 7.9.2) is to type the date of the previous Sunday into an Excel® spreadsheet, using (say) cell A1. Then, type the formula = (A1 − 29226)/7 into another cell—this will show the GPS week which started on that Sunday.

---

* Prior to this definition, the length of a second was defined as 1/86,400 of the time taken for the earth to rotate once with respect to the sun, averaged over the year—hence the term Greenwich *mean* time. Although this avoided the problem of leap seconds, it meant that no-one knew exactly how long a second was, until the end of the year!

7.8.4 Galileo

Galileo is conceptually quite similar to GPS, and also works by measuring the time taken for signals to travel from a satellite to a receiver. However it broadcasts signals on three different frequency bands (1164–1215 MHz, 1260–1300 MHz and 1559–1591 MHz) which should significantly improve the calculation of atmospheric delays. In addition, some of the signals incorporate an integrity check, intended to guard against false indications of position.

As it becomes fully operational,[*] Galileo will bring two major benefits to the surveying community:

1. It will improve accuracy and reduce observation times by providing more satellites: GDOPs of greater than 6 will cease to occur.
2. It should perform noticeably better than GPS in built-up areas, due to the integrity checks in the signals.

On the other hand, Galileo is planned as a commercial venture, in which the users will pay for the deployment and maintenance of the system. In particular, it is intended that commercial users (such as surveyors) should pay for the enhanced positioning services that they will need, by means of access-protection keys on their receivers.

7.9 ENHANCEMENT OF GNSS

7.9.1 Overlay Systems

To enhance the accuracy of GNSS in certain areas, several countries have already developed regional augmentation to the GPS signals, using geostationary satellites. These include WAAS (Wide Area Augmentation System) in the United States, MSAS in Japan, GAGAN in India, and EGNOS (European Geostationary Navigation Overlay System) in Europe. The satellites in these systems firstly improve accuracy by transmitting navigational signals just like any other GPS satellite; this means that receivers can generally 'see' one or more extra satellites in addition to the normal constellation, which reduces the likelihood of periods when the number of visible satellites is too low (or the GDOP is too high) for reliable observations. In addition, these geostationary satellites broadcast more recent, and therefore more accurate, ephemeris data for the constellation—and they also rapidly send out warnings about any satellite which is suspected of giving erroneous data. These features make such augmentation systems

---

[*] With four operational satellites in orbit, Galileo can now (just) be used as an independent system—but even when all four satellites are visible, the GDOP is often high.

very useful (and perhaps essential) in allowing GNSS to be used for certain safety-critical purposes, such as providing landing aids for aircraft.

7.9.2 Precise Ephemerides

The ephemeris data transmitted by each satellite is not a report of its actual orbit, but a prediction of its expected orbit, made up to six hours in advance. The accuracy of satellite surveys is clearly dependent on the accuracy to which the satellites' positions are known—so for the highest accuracy, the satellite observations are recorded and subsequently processed in conjunction with so-called precise ephemeris data, created by observing the satellite's actual orbit and publishing the most accurate results about 12 days later. Precise ephemeris data for GPS satellite orbits is available from the international GNSS, and can be downloaded (March 2013) from http://igscb.jpl.nasa.gov/components/prods_cb.html.

Chapter 8

Geoids, Ellipsoids and Co-ordinate Transforms

8.1 DEFINITION OF THE GEOID

The purpose of engineering surveying is to establish the relative positions of points in three dimensions, so a three-dimensional co-ordinate system is needed to record the findings. The one absolute direction which can always be found is 'up', using nothing more sophisticated than a plumb-bob. For this reason alone, it makes sense to define one axis of the co-ordinate system in the upwards direction. The other two axes can then be conveniently defined to form an orthogonal set with the first axis, and can be thought of as lying at right angles to each other on the surface of a bowl of water held at the foot of the plumb-bob.

Extended over a hundred metres or so, this water surface is virtually a flat plane*—so a simple Cartesian co-ordinate system, with the z-axis pointing upwards, is ideal for small-scale work. When extended part or all of the way round the world, however, the surface (which is actually a surface of constant gravitational potential energy) is not flat, and 'up' is not always in the same direction—so a more elaborate co-ordinate system is needed for larger surveys.

There is an infinite number of surfaces of constant gravitational potential around the earth, each at a different height. The one used by surveyors to define zero height is called the geoid, and can be thought of as the surface defined by the sea level around the world, in the absence of wind or tidal effects.

The geoid is thus straightforward to define, but its exact position in any part of the world can be quite hard to find. In Great Britain, a local realisation of the geoid was created by observing the mean tide height at Newlyn in Cornwall between 1915 and 1921, and establishing a physical datum mark at the resulting average point†. All so-called orthometric heights in Great Britain are measured with respect to this datum. In fact, the shape

---

* Even over this distance, the direction of the vertical will alter by about 3 seconds.
† This is called Ordnance Datum Newlyn, or ODN.

and position of the geoid is steadily changing, due to effects such as continental drift and global warming. Thus, all countries tend in practice to define their own geoid by measuring heights from a fixed datum close to the geoid, rather than using a common geoid.

8.2 THE NEED FOR AN ELLIPSOID

The geoid forms a potentially useful basis for a global co-ordinate system, in that it provides a surface which is exactly horizontal at any point around the earth. If a co-ordinate system could be set up having one axis perpendicular to the surface of the geoid, and two others lying in the surface, then any point on or near the earth's surface could be specified in terms of co-ordinates which identify its horizontal position and its height.

Unfortunately, the exact shape of the geoid is complex and irregular. It dips in the middle of deep oceans, and rises in mountainous areas. However, it does correspond fairly closely (to within 100 metres or so) to a biaxial ellipsoid*—the three-dimensional shape achieved by rotating an ellipse about its minor axis. Taking a suitable ellipsoid to define the basic shape of the earth allows us to define a single co-ordinate system in which we can model the concept of a horizontal plane and a vertical axis anywhere on the earth's surface. This system is known as the *geodetic co-ordinate system* or sometimes as the geographic co-ordinate system; in it, the positions of points on or near the earth's surface are defined in terms of geodetic latitude (ϕ), longitude (λ), and height (h) above an earth-shaped ellipsoid, as shown for point P in Figure 8.1. Varying ϕ and/or λ allows movement on a surface which is always (very nearly) horizontal, while varying h causes movement in a near-vertical direction.

Note that:

1. The line PR is the line passing through P which is perpendicular to the surface of the ellipsoid at the place where it passes through it. It does not, in general, pass through the centre of the ellipsoid.
2. When the size of the ellipsoid's major and minor axes (a and b in Figure 8.1) are known, the geodetic co-ordinates of P (ϕ, λ, h) can be converted into (x, y, z) co-ordinates in a Cartesian system having its origin at the centre of the ellipsoid, the z-axis lying along the minor axis of the ellipse, and the x-axis passing through the zero (Greenwich) meridian. Section 8.5 explains how (and why) this is done.
3. The distance h is called the ellipsoidal height. It is not the same as the orthometric height described in Section 8.1, since no ellipsoid can match the geoid exactly.

---

* Sometimes also called an oblate spheroid in the literature.

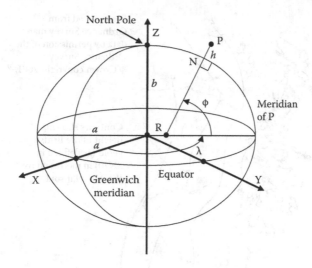

Figure 8.1 Cartesian and geodetic co-ordinates.

Historically, there was no obvious reason for every country to use the same ellipsoid for mapping purposes—so different countries each tended to choose an ellipsoid whose surface corresponded closely to the geoid within their area of interest. These local ellipsoids all have slightly different shapes, and different positions and orientations with respect to the earth's crust. Moreover, since the position of each ellipsoid tends to be defined in terms of fixed points in the country which adopted it, the relationship between them is continually changing, because of continental drift.

In Great Britain, the Airy 1830 ellipsoid was adopted for mapping purposes—this corresponds quite closely to the geoid over the British Isles, being about 1 metre above it along much of the east coast, and about 3 to 4 metres below it along the west coast, as shown in Figure 8.2.

With the advent of GNSS, all countries are now tending to adopt the WGS84 ellipsoid for mapping. This gives the best fit with the geoid over the earth as a whole, but is inevitably less good than the 'national' ellipsoid for individual countries. In the UK, the separation between the geoid and the ETRS89 ellipsoid (an image of the WGS84 ellipsoid that was fixed relative to the UK in 1989) varies between 45 metres and 56 metres across the country, as shown in Figure 8.3. (Note that the geoid has been defined in slightly different ways in Figures 8.2 and 8.3; this difference is negligible, however, compared to the difference between the two ellipsoids.)

The fact that the surface of the geoid is not parallel to any particular ellipsoid gives rise to two effects which need to be considered:

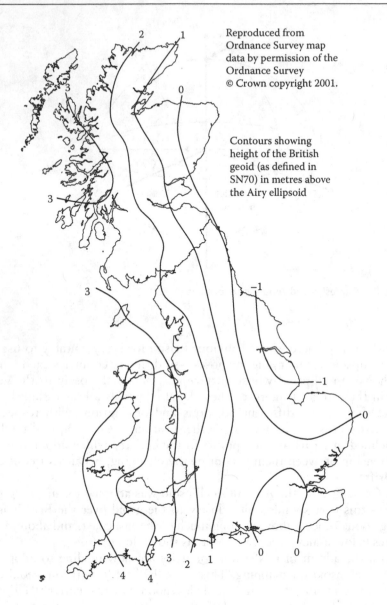

Reproduced from
Ordnance Survey map
data by permission of the
Ordnance Survey
© Crown copyright 2001.

Contours showing
height of the British
geoid (as defined in
SN70) in metres above
the Airy ellipsoid

Figure 8.2 The British geoid on the Airy ellipsoid.

1. If a short plumb-bob was hung from the point *P* in Figure 8.1, it
   would not lie exactly along the line *PN*. Moreover, the neighbouring surfaces of constant gravitational potential are not even exactly
   parallel to each other—so if the string of the plumb-bob was made

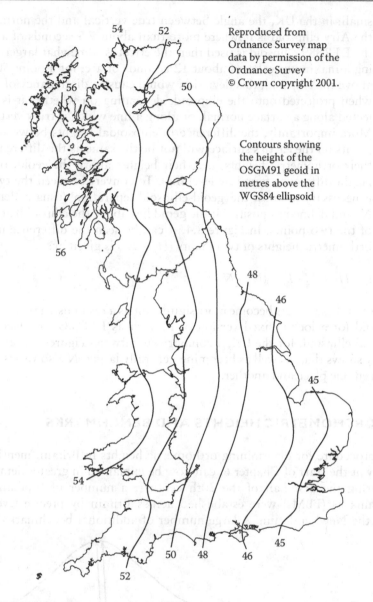

Reproduced from
Ordnance Survey map
data by permission of the
Ordnance Survey
© Crown copyright 2001.

Contours showing
the height of the
OSGM91 geoid in
metres above the
WGS84 ellipsoid

Figure 8.3 The British geoid on the WGS84 ellipsoid.

longer, it might point in yet another direction.* This discrepancy is
known as the deviation of the vertical, and its effect is generally fairly

---

* Assuming it is light compared to the plumb -bob, the string will always show the direction
of 'up' (i.e., the direction normal to the gravitational equipotential surface) at the position
of the plumb-bob.

small: in the UK, the angle between true vertical and the normal to the Airy ellipsoid is nowhere more than about 7.5 seconds of arc. If the ETRS89 ellipsoid is used then the effect is somewhat larger, having a maximum value of about 12 seconds of arc; for a point 500 m above the ellipsoid, this deviation would cause a discrepancy of 3 cm when projected onto the ellipsoid, depending on whether it is projected along a surface normal, or along a line which is truly vertical.

2. More importantly, the difference in ellipsoidal heights between two points on the earth's surface will not be the same as the difference in their orthometric heights (i.e., their heights above the geoid, or the height difference found by levelling). To convert between the two, it is necessary to know the 'geoid-ellipsoid separation' (usually denoted N, and defined as positive if the geoid lies above the ellipsoid) at each of the two points. In Figure 8.4, it can be seen the difference in the orthometric heights of two points $(H_2 - H_1)$ is given by:

$$H_2 - H_1 = (h_2 - h_1) - (N_2 - N_1) \qquad\qquad 8.1$$

Both of these effects become more significant as a result using the WGS84 ellipsoid (or a locally-fixed version of it, such as ETRS89) in place of a national ellipsoid; for the UK, a comparison between Figures 8.2 and 8.3 clearly shows that, as well as becoming generally larger, N also varies more between one place and another.

8.3 ORTHOMETRIC HEIGHTS AND BENCHMARKS

The procedure for determining orthometric heights in Britain, mentioned briefly at the start of Chapter 6, can now be explained in greater detail.

During the first half of the 20th century, a number of fundamental benchmarks (FBMs) were established across Britain by precise levelling from the Newlyn datum. A large number of additional benchmarks were

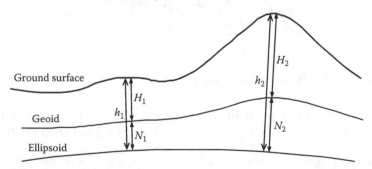

Figure 8.4 Difference between orthometric and ellipsoidal heights.

subsequently established, using the nearby FBMs as reference heights. To find the orthometric height of any point in the UK, a surveyor would simply level from the nearest BM, as described in Chapter 6. Effectively, the shape of the British geoid was defined (or realised) by the physical positions and published heights of those benchmarks.

FBMs consist of elaborately-constructed underground chambers in geologically stable places, containing the actual benchmark. These are topped off with a small pillar which protrudes just above the ground and which carries a domed brass marker, whose height is published and upon which a levelling staff can be placed. The top mark is then normally used for levelling—but if it is destroyed, it can be rebuilt and its height re-established from the FBM below. By contrast, ordinary BMs are cut into walls of buildings, etc., and are much more liable to subsidence and destruction.

The task of regularly checking all BMs throughout the UK and publishing any changes in their heights was a massive and expensive one, and has now been abandoned by the Ordnance Survey because of the advent of GNSS. The positions and heights of all BMs in the UK are still published on the Ordnance Survey's website, but they have not been checked for many years and must now be used with increasing caution, especially in areas liable to subsidence.

Instead of defining the British geoid through a network of BMs, the Ordnance Survey now does so by means of a numerical model called OSGM02.[*] This was constructed by comparing the orthometric heights of 179 FBMs around the country (measured as described above) with their ellipsoidal heights in the ETRS89 system (measured by differential GNSS) to give known values for the separation between the British geoid and the ETRS89 ellipsoid at various places around Britain. These values were incorporated into an interpolation algorithm which then gives an estimated value for the geoid–ellipsoid separation at any other place in the country.[†]

This change has been made possible by the fact that an ellipsoidal height can now be found anywhere in the world to within about 1 cm, using differential GNSS. One receiver is placed at the point whose height is required; another is placed on a GNSS 'passive' station with known co-ordinates, or data are downloaded from a set of nearby 'active' stations. After recording data for about 24 hours (necessary because of the comparative inaccuracy of height readings from GNSS), the co-ordinates of the new station (including its ellipsoidal height) are computed. In the UK, the OSGM02 model then provides the value of N (Figure 8.4) and hence the orthometric height of the station.

---

[*] This supplants OSGM91, used for the production of Figure 8.3.

[†] Note that this method does not necessarily provide the exact shape of the true geoid passing through the Newlyn datum, as the FBMs may have changed their heights since they were measured, due to subsidence and land uplift. The inaccuracies inherent in the definition of OSGM02 are, however, probably small by comparison with the errors inherent in making practical use of it.

To find *differences* in heights over a short distance, it is more accurate and often quicker to use levelling or trigonometric heighting techniques (fixing one height by GNSS and a geoid model if absolute orthometric heights are also required) than to fix both heights independently by GNSS. However, the best accuracy which can be expected even from precise levelling is about $2\sqrt{d}$ millimetres, where d is the distance between the two points in kilometres; if the two points are more than 25 km apart, it is probably more accurate (and certainly much quicker!) to find their relative heights by means of two separate GNSS observations.

8.4 GEOMETRY OF THE ELLIPSE

It is not the intention of this book to explain all aspects of ellipses, but simply those which are necessary for a proper understanding of geodetic surveying.

8.4.1 Defining the Shape of an Ellipse

The shape of an ellipse is most simply defined by the size of its major and minor semi-axes, which are generally labelled a and b, as in Figure 8.5. In geodesy, ellipses are generally nearly circular, and their shapes are therefore defined in terms of their major semi-axis and either their flattening (f) or their reciprocal of flattening (r), where

$$r = \frac{1}{f} = \frac{a}{a-b} \qquad\qquad 8.2$$

whence

$$b = a(1-f) = a\left(1 - \frac{1}{r}\right) \qquad\qquad 8.3$$

The formula for the ellipse can then be simply expressed as

$$\left(\frac{x}{a}\right)^2 + \left(\frac{y}{b}\right)^2 = 1 \qquad\qquad 8.4$$

Another useful characteristic of an ellipse is its eccentricity, e, which is defined by

$$e^2 = \frac{a^2 - b^2}{a^2} \quad \text{whence} \quad \frac{b^2}{a^2} = 1 - e^2 \quad \text{or} \quad b = a\sqrt{1-e^2} \qquad\qquad 8.5$$

The value of e can also be obtained directly from the reciprocal of flattening, using the formula

$$e^2 = \frac{2r-1}{r^2}$$ 8.6

There are various ways to define a position on an ellipse. The one mainly used in surveying is the geodetic latitude, ϕ, defined as the point on the ellipse where the surface normal makes an angle of ϕ with the x-axis (or the plane of the equator, if on the surface of the earth) as shown in Figure 8.5. The same point can also be defined by its geocentric latitude, namely the angle between the x-axis and the line ON (omitted from Figure 8.5 for clarity), or by its parametric latitude (θ) as shown in the figure. The latter angle (also known as the reduced latitude or, in the context of orbital theory, the eccentric anomaly) defines a point Q on the auxiliary circle (see Figure 8.5) which lies exactly above the point on the ellipse and serves as a useful parameter for defining the shape of the ellipse, namely,

$$x = a\cos\theta \text{ and } y = b\sin\theta$$ 8.7

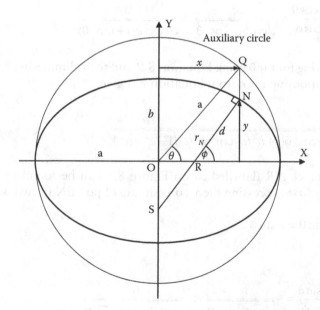

Figure 8.5 Definition of an ellipse.

8.4.2 Curvature on an Ellipse

To find the curvature at a point on an ellipse, we first need the distance NS (S being where the normal through N intersects with the y-axis) in Figure 8.5, which we will call r_N. This can be found as follows.

First, using the relationships in Equation 8.7, the slope of the ellipse at any point can be defined in terms of its parametric latitude:

$$\frac{dy}{dx} = \frac{dy/d\theta}{dx/d\theta} = -\frac{b}{a}\cot\theta \qquad\qquad 8.8$$

But this slope can also be defined as being $(-\cot\phi)$, so we can write

$$\tan\theta = \frac{b}{a}\tan\phi \qquad\qquad 8.9$$

We can also see that the x co-ordinate of point N (which we can simply call x) can be expressed in two ways:

$$x = a\cos\theta = r_N\cos\phi \qquad\qquad 8.10$$

This equation can be re-arranged to give:

$$r_N = \frac{a\cos\theta}{\cos\phi} = \frac{a}{\sqrt{\cos^2\phi/\cos^2\theta}} = \frac{a}{\sqrt{\cos^2\phi(1+\tan^2\theta)}} \qquad\qquad 8.11$$

Substituting for $\tan\theta$ using Equation 8.9 and then eliminating b using the second relationship for e from Equation 8.5 gives:

$$r_N = \frac{a}{\sqrt{\cos^2\phi(1+b^2/a^2\tan^2\phi)}} = \frac{a}{\sqrt{1-e^2\sin^2\phi}} \qquad\qquad 8.12$$

The distance NR (labelled d) on Figure 8.5 can be found in a similar manner, by first expressing the y co-ordinate of point N in two ways:

$$y = b\sin\theta = d\sin\phi$$

whence

$$d = \frac{b\sin\theta}{\sin\phi} = \frac{b}{\sqrt{\sin^2\phi/\sin^2\theta}} = \frac{b}{\sqrt{\sin^2(1+\cot^2\theta+1)}} \qquad\qquad 8.13$$

This can be simplified in a similar way to Equation 8.12:

$$d = \frac{b}{\sqrt{\sin^2(a^2/b^2\cot^2\phi+1)}} = \frac{b\sqrt{1-e^2}}{\sqrt{\sin^2\phi(\cot^2\phi+1-e^2)}} = \frac{a(1-e^2)}{\sqrt{1-e^2\sin^2\phi}} \qquad 8.14$$

and can then be combined with the result of Equation 8.12 to give

$$d = (1-e^2)\, r_N \qquad 8.15$$

To obtain the curvature of the ellipse at N, we start with the standard formula which defines the curvature κ at any point on a line in two dimensions, namely

$$\kappa = \frac{d^2y/dx^2}{\left\{1+(dy/dx)^2\right\}^{3/2}} \qquad 8.16$$

Now

$$\frac{dy}{dx} = -\cot\phi$$

as before, so

$$\frac{d^2y}{dx^2} = \frac{d}{d\phi}\left(\frac{dy}{dx}\right)\frac{d\phi}{dx} = \frac{1}{\sin^2\phi}\frac{d\phi}{dx} = \frac{1}{\sin^2\phi\,\dfrac{dx}{d\phi}} \qquad 8.17$$

But $x = r_N\cos\phi$ so we can express r_N using Equation 8.12 and write

$$\frac{dx}{d\phi} = \frac{d}{d\phi}\left(\frac{a\cos\phi}{(1-e^2\sin^2\phi)^{1/2}}\right) = \frac{a\sin\phi(e^2-1)}{(1-e^2\sin^2\phi)^{3/2}} \qquad 8.18$$

Substituting this result into Equation 8.17, and then putting the expressions for dy/dx and d^2y/dx^2 into Equation 8.16 gives

$$\kappa = \frac{(1-e^2\sin^2\phi)^{3/2}}{a\sin^3\phi(e^2-1)(1+\cot^2\phi)^{3/2}} = \pm\frac{(1-e^2\sin^2\phi)^{3/2}}{a(e^2-1)} \qquad 8.19$$

The ambiguity of sign in Equation 8.19 reflects the fact that positive curvature is (by convention) upwards—so the curvature of our ellipse is negative above the x-axis, and positive below it. We know that, for an ellipse, the curvature is always inwards and e is always less than unity—so the (inwards) radius of curvature can simply be written as

$$r_M = \frac{1}{|\kappa|} = \frac{a(1-e^2)}{(1-e^2 \sin^2 \phi)^{3/2}} \qquad\qquad 8.20$$

which will always yield a positive number. This radius of curvature is conventionally referred to as r_M (or just as M, in some literature) because it becomes the curvature of a meridian when the ellipse is spun around its minor axis to form an ellipsoid.

8.5 TRANSFORMATIONS BETWEEN ELLIPSOIDS

Because points on the earth's surface can be expressed in terms of different Cartesian/geodetic co-ordinate systems, there is often a need to convert a point's co-ordinates from one system to another. This is particularly the case when GNSS information (which is measured in the WGS84 system, or perhaps in a local realisation such as ETRS89) needs to be expressed in terms of a national ellipsoid, such as Airy 1830. The bottom half of Figure 7.3 shows how this conversion forms a key stage in the processing of GNSS observations.

The central step of the conversion is done in Cartesian space, using a transform matrix. Such matrices are always conformal, meaning that they preserve the *shape* of a set of points during the transform, but they sometimes have a scale factor incorporated, to allow the *size* of the set to change.

Before it can be used, a transform matrix has to be defined. It is easier to understand the definition process once the method of use is understood, so the two topics are explained in that order.

8.5.1 Using a Transform Matrix

Typically, a set of geodetic co-ordinates in one system (system A, say) need to be converted to another system (system B). The process for doing this is as follows.

First, the (ϕ, λ, h) (i.e., latitude, longitude, ellipsoidal height) co-ordinates of a point in system A are converted into the corresponding (x, y, z) co-ordinates, as set out in Figure 8.1 This requires the shape parameters of the ellipse in question which, as explained above, are normally given in terms of the semi-major axis and the reciprocal of flattening. The eccentricity of the ellipse can then be obtained using Equation 8.6.

The distance NS in Figure 8.6 is r_N, which is given by Equation 8.12; and the distance NR is d, which is given by Equation 8.14. The Cartesian co-ordinates of P are then easily obtained by the following formulae:

$$x = (r_N + h)\cos\phi\cos\lambda \qquad\qquad 8.21$$

$$y = (r_N + h)\cos\phi\sin\lambda \qquad\qquad 8.22$$

$$z = (d + h)\sin\phi = \left(r_N\{1 - e^2\} + h\right)\sin\phi \qquad\qquad 8.23$$

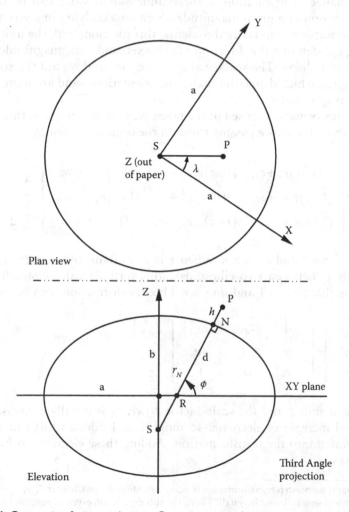

Figure 8.6 Converting from geodetic to Cartesian co-ordinates.

The next step is to transform the (x,y,z) co-ordinates of P in system A into corresponding co-ordinates in system B, which we will term (x',y',z'). This transformation consists of a translation and rotation to accommodate the fact that the two ellipsoids have different centres and orientations. Sometimes, though, a scale factor is also defined as part of the transform. This is fundamentally illogical, since 1 metre in system A should equal 1 metre in system B, albeit in a different orientation. The purpose, however, is to allow for the fact that the sets of fixed markers which still define some ellipsoids are not the exact distance apart that they have been defined to be, due to the difficulty of measuring accurate distances at the time when the co-ordinates of the markers were established.

Any change in orientation in three-dimensional space can be thought of as a rotation of a given magnitude about an axis lying in a given direction. The vector $r\underline{e}$ can be used to define this rotation, with the unit vector $\underline{e} = [e_x e_y e_z]^T$ defining the direction of the axis, and r the magnitude of the rotation in radians. The scalar values of re_x, re_y and re_z are the so-called Euler angles, which define the change in orientation—and are more simply written as r_x, r_y and r_z.

The transformation caused to the point (x, y, z) by a rotation through an angle r about the axis e passing through the origin, is given by:

$$
\begin{bmatrix} x' \\ y' \\ z' \end{bmatrix} = \begin{bmatrix} (1-c)e_x^2+c & (1-c)e_xe_y-se_z & (1-c)e_xe_z+se_y \\ (1-c)e_xe_y+se_z & (1-c)e_y^2+c & (1-c)e_ye_z-se_x \\ (1-c)e_xe_z-se_y & (1-c)e_ye_z+se_x & (1-c)e_z^2+c \end{bmatrix} \begin{bmatrix} x \\ y \\ z \end{bmatrix} \qquad 8.24
$$

where $c = \cos r$ and $s = \sin r$. When r is in radians and is very small (as it usually is between two ellipsoids),[*] this formula can be simplified by assuming that $\cos r = 1$ and $\sin r = r$. The transformation then becomes:

$$
\begin{bmatrix} x' \\ y' \\ z' \end{bmatrix} = \begin{bmatrix} 1 & -re_z & re_y \\ re_z & 1 & -re_x \\ -re_y & re_x & 1 \end{bmatrix} \begin{bmatrix} x \\ y \\ z \end{bmatrix} \qquad 8.25
$$

The origin shift $\underline{t}$ and the scale factor s (which is usually expressed as a fractional increase or decrease, so must have 1 added to it) can now be incorporated into the transformation. Adding these elements to Equation 8.25 gives

---

[*] Note that if a one-step transformation is being established (see Chapter 7, Section 7.6) the rotation will generally not be small. Then, the full expression given in equation 8.24 must be used.

$$\begin{bmatrix} x' \\ y' \\ z' \end{bmatrix} = \begin{bmatrix} t_x \\ t_y \\ t_z \end{bmatrix} + \begin{bmatrix} 1+s & -r_z & r_y \\ r_z & 1+s & -r_x \\ -r_y & r_x & 1+s \end{bmatrix} \begin{bmatrix} x \\ y \\ z \end{bmatrix} \qquad 8.26$$

It is important to realise that Equation 8.26 is not actually being used to move the point at (x, y, z)! Rather, it is being used to re-express the point's co-ordinates in terms of a different (x', y', z') set of axes, which are not exactly co-incident with the original (x, y, z) axes.

Finally, the transformed (x',y',z') co-ordinates are converted back into (ϕ',λ',h') co-ordinates on ellipsoid B. Inspection of Equations 8.21 to 8.23 will show that this is not as straightforward as converting from (ϕ,λ,h) to (x,y,z), but it can be done as follows.

The value of λ' can be found by dividing Equation 8.22 by Equation 8.21 and re-arranging to give:

$$\lambda' = \tan^{-1}\left(\frac{y'}{x'}\right) \qquad 8.27$$

remembering that, if x' is negative, then λ' lies between 90° and 270° west of Greenwich.

ϕ' is most easily found by iteration. Equations 8.21 and 8.22 can be combined to give

$$(r_N' + h')\cos\phi' = \sqrt{x'^2 + y'^2} \qquad 8.28$$

Since h is generally much smaller than r_N, a good first guess for ϕ' is to assume that h' is zero, and use Equation 8.28 to write:

$$r_N' \cos\phi' \approx \sqrt{x'^2 + y'^2} \qquad 8.29$$

Likewise, we use Equation 8.23 with h' set to zero, to obtain

$$r_N' \sin\phi' \approx \frac{z'}{1-e'^2} \qquad 8.30$$

Dividing Equation 8.30 by Equation 8.29 gives a first guess for ϕ':

$$\phi_1' = \tan^{-1}\left(\frac{z'}{(1-e'^2)\sqrt{x'^2+y'^2}}\right) \qquad 8.31$$

This value is then used in Equation 8.12 to obtain a first guess for r'_N (with $n = 1$):

$$r'_{Nn} = \frac{a'}{\sqrt{1 - e'^2 \sin^2 \phi'_n}} \qquad\qquad 8.32$$

Equation 8.23 can now be rewritten as:

$$(r'_N + h') \sin \phi' = z' + r'_N e'^2 \sin \phi' \qquad\qquad 8.33$$

and combined with Equation 8.28 to obtain an improved guess for ϕ':

$$\phi'_{n+1} = \tan^{-1} \left(\frac{z' + r'_{Nn} e'^2 \sin \phi'_n}{\sqrt{x'^2 + y'^2}} \right) \qquad\qquad 8.34$$

Equations 8.32 and 8.34 are now used repeatedly to get improved values of r_N' and ϕ' respectively, until the changes in ϕ' are within the accuracy required. Finally, h' is obtained from Equation 8.28, using the final values for r'_N and ϕ':

$$h' = \frac{\sqrt{x'^2 + y'^2}}{\cos \phi'} - r'_N \qquad\qquad 8.35$$

A worked example of this process is given in Appendix C. It is worth noting that there are several other ways of converting from Cartesian to geodetic co-ordinates, some of which are arguably more efficient than the method shown here. However, this method is reliable, and relatively easy to understand.

The following points are useful when applying the transform:

1. Be sure to use the given parameters correctly in Equation 8.26. The values of r_x, r_y and r_z must be in radians, and the value of $(1 + s)$ should be very close to unity. (The given angles are often quoted in seconds of arc, and the scale factor is often quoted in parts per million.)
2. Remember that if (say) millimetre precision is required from the conversion, this is about one part in 10^{10}, compared with the (x,y,z) co-ordinate of the point. All calculations must therefore be done to 10 significant figures, and 'double precision arithmetic' should be used in computer programs.
3. It is useful to have at least one point whose position is already known in both systems to test that the transform has been set up correctly before further points are transformed.

4. A reverse transform can be obtained by changing the signs of the shifts, rotations and scale factor quoted for the forward transform and using Equation 8.26. Note that this reverse transform is not the exact inverse of the original transform matrix, so any transformed points will not return exactly to their original positions; the resulting position errors will be of the order of δ^2/R_E, where δ is the movement caused by the rotational part of the transform and R_E is the earth's radius $(6.37 \times 10^6$ m). This is because Equation 8.26 is based on the approximate version of the transform matrix shown in Equation 8.25, rather than the exact (and conformal) version given in Equation 8.24. The resulting errors are usually negligible, but might become significant for Euler angles larger than 1 second.

8.5.2 Making a Transform Matrix

The process of making a transform matrix is computationally similar to least-squares adjustment (see Chapter 11) and several computer programs offer the necessary functionality, so it need not be explained in detail here. A brief overview is, however, useful for surveyors.

The process starts with an initial set of three or more points with Cartesian co-ordinates in two different systems (A and B), and involves adjusting the seven variables in Equation 8.26 so as to map all the points in the first system as closely as possible onto their counterparts in the second system.

An exact match can only be achieved if the relative positions of the points are identical in both systems—and this is not usually the case if those relative positions were surveyed in different ways, in the two systems. The goal is to minimise the sum of the squares of the distances between each transformed point from system A and its counterpart in system B, and this becomes easier if the scale factor (s, in Equation 8.26) is allowed to vary. Sometimes, though, it is desirable to constrain this to be zero—this ensures that the distance measurements made in system A (which may be much more accurate than those in system B) will be preserved during the transform process. If system B has the more accurate distances, then it would first be necessary to find the best transform from system B to system A, and then invert the transform as described at the end of Section 8.5.1.

A minimum of three points (in each system) is required to set up the transform, for the same reason that a table needs a minimum of three legs for stability. If more points are used, the robustness of the process will be improved, as it will become possible to detect any points which have been incorrectly specified in one system or the other. Following the 'table' analogy above, it is also important that the points are well spread out and do not lie in a near-straight line, as this will make the transform unreliable if it is applied to further points which do not lie close to the same line. This is a problem which is particularly likely to be encountered by road or rail engineers, whose site may be long and straight.

Quite often, the two sets of co-ordinates will not be in Cartesian form—they may be geodetic co-ordinates or grid co-ordinates (as discussed in Chapter 9). If so, then the computer program will also require the necessary projection details to convert the grid co-ordinates to geodetic form, and the parameters of the ellipsoid to convert the geodetic co-ordinates to Cartesian form. (The steps in this conversion process are shown in Figure 7.3.)

Having set up such a transform, it is important to inspect the results returned by the computer program. In particular, the residual errors should be checked to see whether there are any outlier points which may have been wrongly measured or specified, and which may be adversely affecting the transform. If many points are available to define the transform, it is instructive to split them into two groups, generate a transform from each group, and then compare the two transforms.

8.6 ETRS89 AND THE INTERNATIONAL TERRESTRIAL REFERENCE SYSTEM

A common application of ellipsoid transformations is the conversion of co-ordinates between the physical realisation of WGS84 on a particular continent (done via ETRS89 in Europe) and the International Terrestrial Reference Frame (ITRF), which now provides the definitive realisation of WGS84 on the earth's surface through measurement of the WGS84 co-ordinates of a number of 'fixed' points around the world.

As explained in Chapter 7, ETRS89 is based on the WGS84 co-ordinates that a number of European control stations had on 1 January 1989, but it also takes advantage of subsequent (and more accurate) measurements of the relative positions of those stations. At the time of writing, ETRS89 is realised by means of a reference frame called ETRF2000, which consists of the ground stations themselves plus the current best estimates of where those stations were in the WGS84 system on 1 January 1989. These estimates were made by using the positions and velocities of those stations on 1 January 2000 (as published in ITRF2000) to see where they would have been on 1 January 1989. A best-fit transform was then generated between ITRF2000 and ETRF2000, as described in the previous section, and this can be used to convert the ITRF2000 co-ordinates of any other point into ETRS89 co-ordinates. The parameters of this transform are given in Appendix A. Note that these parameters include a time element since ITRF2000 co-ordinates change with time, as defined by the published velocities of its ground stations. Thus, if the ITRF2000 co-ordinate of a point is established on, say, 1 October 2014, the value of t in Appendix A would be set to 2014.75 to find its co-ordinates in ETRS89.

Since 2000, though, two further ITRFs have been published—the latest one is ITRF2008, whose published positions and velocities will be used to estimate the WGS84 co-ordinates of control stations around the world until the publication of the next ITRF. This includes the control stations that are used to observe the constellation of GNSS satellites, from which 'precise ephemeris' data is computed.

The most accurate GNSS observations are processed using precise ephemeris data, and the most accurate DGNSS makes use of the ITRF co-ordinates of any base stations*; so the results are always in terms of the current ITRF (i.e., no longer in terms of ITRF2000). Fortunately, each new ITRF is provided with a set of associated transforms which allow conversion to the earlier ITRFs. These transforms are also time-dependent, to allow for the slightly different station velocities given in the different ITRFs. In particular, ITRF2008 provides a transform to ITRF2000, whose parameters are shown in Appendix A.

The process for generating ETRS89 co-ordinates from GNSS observations using precise ephemeris data is thus a two-stage one, at present. First, the ITRF2008 to ITRF2000 transform is used to convert the results to ITRF2000 co-ordinates, with the time (t) set to the average time at which the observations were made. Then, the ITRF2000 to ETRS89 transform is applied, with t once again set to the time of the observations.

8.7 FURTHER PROPERTIES OF ELLIPSOIDS

A few more properties of ellipsoids now also need to be explained, as they will be needed in the next chapter.

8.7.1 Curvature on an Ellipsoid

8.7.1.1 Principal Curvatures

At any point on a smoothly curved surface, a tangent plane and surface normal vector can be found, as shown in Figure 8.7; and if a direction along the tangent plane is then chosen, a normal plane is defined which contains the point, the chosen direction, and the surface normal. A path on the surface is also defined, where this normal plane cuts through the surface.

---

* ETRS89 co-ordinates are becoming compromised by the fact that its stations are all moving *around* the earth, rather than in parallel straight lines over its surface—and the distances they have moved since 1989 mean that the difference between ETRS89 and WGS84 is now starting to show a rotational element in addition to simple translation. This does not matter for most surveying purposes, but it makes ETRS89 unsuitable for very high accuracy surveying work within Europe—and ETRS89 is in any case inappropriate for work which extends to other tectonic plates.

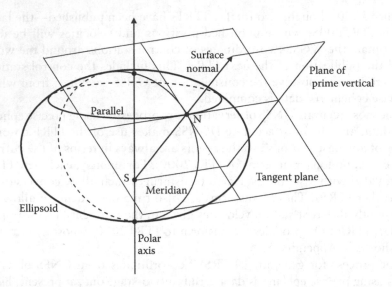

Figure 8.7 Tangent and prime vertical planes on an ellipsoid.

If a second surface normal vector is established a short distance along such a path, it will in general not lie in the normal plane used to make the path.* There are always two directions, however, where this second surface normal does lie in the normal plane, i.e., where there is no *torsion* along the path; these two directions are called the *principal directions*, and the local curvatures of the associated paths are called the *principal curvatures* at the point.† In general one of these two paths will have a higher curvature than the other—and paths on all other normal planes through the point will have local curvatures whose magnitude lies somewhere between those two limits, plus some degree of torsion.

In the special case of a sphere, the curvature of every direction through a point is the same, and any two orthogonal directions can be taken as the principal directions. On an ellipsoid, however, the curvature varies (except at the poles): one of the principal directions always lies along the meridian which passes through the point; the other principal direction is perpendicular to that, and its associated normal plane (which is orthogonal to the tangent plane and to the plane of the meridian) is called the prime vertical. These are shown in Figure 8.7.

---

* This is because the orientation of the surface twists with respect to the direction of the path, a behaviour which is known as geodetic torsion.
† The proof of this was published by Leonard Euler in 1760, and it is known as Euler's theorem.

We have already (Equation 8.20) derived an expression for the radius of curvature of the meridian running through a point on the surface of an ellipsoid, so we can write:

$$\kappa_M = \frac{1}{r_M} = \frac{(1 - e^2 \sin^2 \phi)^{3/2}}{a(e^2 - 1)}$$ 8.36

The curvature of the path in the prime vertical can also be found quite easily. One formulation of Meusnier's theorem[*] states that

$$\kappa_N = \kappa_I \cos \alpha$$ 8.37

where κ_N is the curvature of a surface path lying in a normal plane, κ_I is the curvature of the surface path lying an inclined plane (containing the point and the chosen direction, but not the surface normal), and α is the angle between these two planes.

Looking at Figure 8.8, and taking the prime vertical at N as the normal plane, we can see that a suitable inclined plane is the one which contains the parallel passing through N, known as the parallel of latitude. The surface path lying in this inclined plane is of course the parallel itself, which is a circle of radius $r_N \cos \phi$. The surface path therefore has a curvature of

$$\frac{1}{r_N \cos \phi} .$$

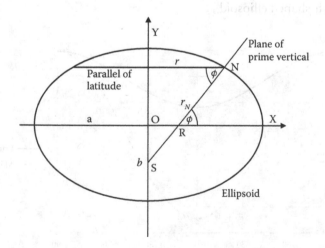

Figure 8.8 **Prime vertical and parallel of latitude**

---

[*] Named after the French mathematician Jean-Baptiste Meusnier, who proved it in 1779.

We can also see that the parallel of latitude is inclined to the prime vertical at an angle ϕ. Applying Equation 8.37 therefore gives

$$\kappa_N = \kappa_I \cos\phi = \frac{\cos\phi}{r_N \cos\phi} = \frac{1}{r_N} \qquad\qquad 8.38$$

which (very conveniently) shows that the radius of curvature in the prime vertical is r_N. We can therefore use Equation 8.12 and write:

$$\kappa_N = \frac{\sqrt{1 - e^2 \sin^2\phi}}{a} \qquad\qquad 8.39$$

We now have formulae to compute the radius of curvature along the meridian and along the prime vertical at any latitude. Figure 8.9 shows how these principal radii of curvature vary with geodetic latitude on the WGS84 ellipsoid.

8.7.1.2 Curvature in Other Directions

Once the magnitudes and directions of the two principal curvatures have been found at a point, the curvature along any other direction is also defined. A straightforward way of finding these is to use the Mohr's circle construction, which is traditionally used to calculate plane and shear stress in nonprincipal directions. The construction for this is shown in Figure 8.10; in addition to calculating the curvature along any path, it can also calculate the torsion (known as the *geodetic torsion* when computed on an earth-shaped ellipsoid).

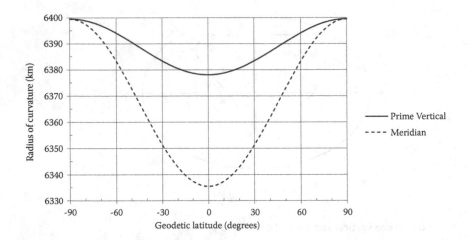

Figure 8.9 Principal radii of curvature on the WGS84 ellipsoid.

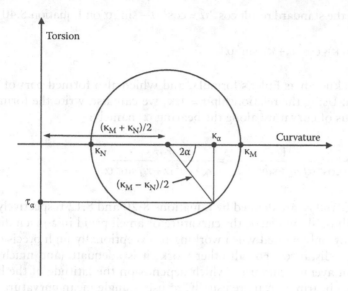

Figure 8.10 Mohr's circle construction for curvature and torsion.

As can be seen from Figure 8.9, the (convex) curvature along the meridian is always higher than that along the prime vertical, so the principal curvatures will always appear along the horizontal axis of the diagram in the order shown in Figure 8.10. We first draw a circle centred on the axis of curvature which passes through these two points, and then travel round it clockwise by an angle of 2α from the reference direction (i.e., northwards along the meridian). The x and y co-ordinates of the point thus defined give the curvature and torsion of the surface along a direction at an angle α clockwise from the reference direction, i.e., along a path having a bearing (or azimuth) of α with respect to true north.

By inspection of the diagram, the circle is centred at $(\kappa_M + \kappa_N)/2$, and has a radius of $(\kappa_M - \kappa_N)/2$. The formulae for the curvature and torsion along the line with bearing α can therefore be expressed as:

$$\kappa_\alpha = \frac{\kappa_M + \kappa_N}{2} + \left(\frac{\kappa_M - \kappa_N}{2}\right)\cos 2\alpha = \frac{\kappa_M(1 + \cos 2\alpha)}{2} + \frac{\kappa_N(1 - \cos 2\alpha)}{2} \quad 8.40$$

and

$$\tau_\alpha = -\left(\frac{\kappa_M - \kappa_N}{2}\right)\sin 2\alpha = (\kappa_N - \kappa_M)\sin\alpha\cos\alpha \quad 8.41$$

where κ_M and κ_N are defined in Equations 8.36 and 8.39 respectively and positive torsion implies clockwise rotation as viewed in the direction of travel, i.e., a 'right-hand thread'.

Using the standard result $\cos 2\alpha = \cos^2 \alpha - \sin^2 \alpha$ on Equation 8.40 gives:

$$\kappa_\alpha = \kappa_M \cos^2 \alpha + \kappa_N \sin^2 \alpha \qquad 8.42$$

which is known as Euler's formula, and which also formed part of Euler's theorem. Using the relationship $r = 1/\kappa$, we can also write the formula for the radius of curvature along the bearing α, namely:

$$r_\alpha = \frac{1}{\cos^2 \alpha / r_M + \sin^2 \alpha / r_N} = \frac{r_M r_N}{r_N \cos^2 \alpha + r_M \sin^2 \alpha} \qquad 8.43$$

where r_M and r_N are defined by Equations 8.20 and 8.12 respectively.

As will be shown later, the curvature of an ellipsoid in a given direction is actually only needed when working to exceptionally high precision over very long distances. For all other work, it is adequate (and much easier) to use an average curvature which depends on the latitude of the line but not on its bearing; or, more usually, to use a single mean curvature for the entire ellipsoid. Methods for calculating both of these are given below.

8.7.1.3 Average Curvature at a Point

There are two possible approaches to finding an average radius of curvature at a point on the ellipsoid. One is to find the average curvature of all the paths running through the point: this can be seen to be $(\kappa_M + \kappa_N)/2$ by inspection of Figure 8.10, or can be formally evaluated as follows:

$$\kappa_{AV} = \frac{1}{2\pi} \int_0^{2\pi} \kappa_\alpha d\alpha = \frac{1}{2\pi} \int_0^{2\pi} (\kappa_M \cos^2 \alpha + \kappa_N \sin^2 \alpha) d\alpha = \frac{\kappa_M + \kappa_N}{2} \qquad 8.44$$

Then, an average radius of curvature can be evaluated:

$$r_{AV1} = \frac{2}{1/r_M + 1/r_N} = \frac{2 r_M r_N}{r_M + r_N} \qquad 8.45$$

where r_M and r_N are defined by Equations 8.20 and 8.12 respectively.

The second approach is to evaluate an average radius of curvature directly, using Equation 8.43:

$$r_{AV2} = \frac{1}{2\pi} \int_0^{2\pi} r_\alpha d\alpha = \frac{1}{2\pi} \int_0^{2\pi} \frac{r_M r_N}{r_M \cos^2 \alpha + r_N \sin^2 \alpha} d\alpha = \sqrt{r_M r_N} \qquad 8.46$$

This second average radius is the reciprocal of the geometric mean of the principal curvatures, whereas the first one is the reciprocal of their arithmetic mean. In the case of an earth-shaped ellipsoid, the difference between these two values is nowhere greater than about 10 parts per million, and the resulting effect on subsequent calculations is several orders of magnitude smaller than this—so the arguments for using one in preference to the other are fairly academic.

8.7.1.4 Mean Curvature of an Ellipsoid

As with the average curvature at a point, there are several possible approaches to specifying a mean radius of curvature for the whole ellipsoid. One obvious approach is to use the radius of the sphere which has the same volume as the ellipsoid:

$$\frac{4\pi}{3} R_E{}^3 = \frac{4\pi}{3} a^2b \text{ whence } R_E = \sqrt[3]{a^2b} \qquad\qquad 8.47$$

where R_E is the mean earth radius, and a and b are the major and minor semi-axes of the ellipsoid. Another approach is the one specified by the International Union of Geodesy and Geophysics, namely

$$R_E = \frac{2a+b}{3} \qquad\qquad 8.48$$

For a near-spherical ellipsoid (such as the earth), the difference between these two approaches is negligible: using the parameters for the WGS84 ellipsoid, they give values of 6,371,001 metres and 6,371,009 metres, respectively.

A third possible approach is to take the arithmetic mean between the maximum curvature on the ellipsoid (along a meridian at the equator) and the minimum value (at either pole, in any direction). The reciprocal of this mean gives a radius of 6,367,353 metres on the WGS84 ellipsoid. This approach is of interest because it comes closer than the others to minimising the largest error that can be caused by using a single average value of R_E in place of the correct value, when calculating a distance over the ellipsoid at any place in the world. However, the improvement by comparison with the two methods described above is not very significant, so 6,371 km can confidently be taken as a suitable mean radius for the earth.

8.7.2 Geodesics

An important concept in surveying is the horizontal distance between two points. In the context of geodesy, this translates into the length of the shortest line over the surface of the ellipsoid, between the places where the two points project onto it.

On an earth-shaped ellipsoid, the mathematical term for such a line is a geodesic. If the ellipsoid was a perfect sphere, the geodesic line would lie along a great circle—a circle with its centre at the centre of the sphere, and passing through the two projected points. Great circle geodesics are relatively easy to compute and analyse, as their radius is constant and they lie in the plane defined by the centre of the sphere and the two endpoints. On an ellipsoid, though, things become more difficult—firstly because the geodesics do not in general lie in planes; and secondly because their lengths cannot be expressed by a single explicit formula but need to be evaluated by numerical integration.

It will be shown in this chapter and the next that a full understanding of geodesics is not often required, even when surveying to high accuracy. However, surveyors should at least have a basic understanding of their nature, so a brief introduction is presented here.

At any point on a curved line in three-dimensional space we can define a direction of travel; by taking two further points a short distance on either side of the first one, we can also define a plane in which the line is curving—since the line is curved, the three points will not lie in a straight line, so the plane is fully defined. If a circle is now drawn in this plane so as to pass through the three points, its centre will be at the local centre of curvature of the line. This circle is known as the *osculating circle* of the curve (from the Latin word 'osculare', to kiss) because it provides the circular arc which gives the closest possible fit to the curve in the region of interest. The plane of the circle is known as the *osculating plane*.

On a geodesic, the osculating plane of the curve is always a normal plane of the surface, as defined in Section 8.7.1 above. The French mathematician Alexis Clairaut (1713–1765) developed a key formula from this insight, which is known as Clairaut's relation:[*]

$$r \sin \alpha = \text{constant} \qquad\qquad 8.49$$

where r is the distance between the point on the geodesic and the polar axis of the ellipsoid, and α is the local bearing (or azimuth angle) of the geodesic. Noting that

$$r = r_N \cos \phi \qquad\qquad 8.50$$

---

[*] A simple explanation of Clairaut's relation is as follows. Imagine a particle P which is free to slide over the surface of the ellipsoid, but constrained to remain on it. The force on P will always act in a direction perpendicular to its movement, so the magnitude of P's velocity (v, say) will stay constant. Also, the line of the force always passes through the axis of the ellipsoid, so P's moment of momentum about the axis (given by $r \times v \sin\alpha$) will stay constant. Since v is constant, this means that $r \sin\alpha$ must also stay constant.

from Figure 8.8, we can use Clairaut's relation and Equation 8.12 to relate the bearing of the geodesic to its latitude:

$$\frac{1}{\sqrt{1-e^2 \sin^2 \phi}} \cos\phi \sin\alpha = \text{constant} \qquad 8.51$$

The 'direct problem' in geodesic literature takes a starting point (ϕ, λ), a starting bearing (α) and a length for the geodesic as 'input', and calculates the resulting endpoint as the 'output'. To solve this, the given values of e, ϕ and α are first used to find the value of the constant in Equation 8.51. A small movement is then made in the direction defined by α, the distance moved and new values of ϕ and λ are computed, and Equation 8.51 is used to find a new value of α. This process is repeated until the geodesic has the required length, and the final position on the ellipsoid can be returned.

A more typical problem in surveying is to find the length of the geodesic between two known points on the ellipsoid, and its bearing at each end. This is known as the 'inverse problem' in the literature, and is slightly harder to solve than the 'direct problem'. The process involves guessing a starting value for α, and solving the direct problem until the geodesic reaches its closest point of approach to the given endpoint. Then, an improved guess for the initial value of α is computed, and the process is repeated until the geodesic passes through (or acceptably close to) the endpoint.

There is a large and still-growing quantity of literature on computing geodesics, as the problem is also of active interest to several scientific fields apart from surveying. The main focus of this literature is to find computationally efficient methods which solve the two problems mentioned above to a high level of accuracy, and the details of this are beyond the scope of this book. One useful (and freely available) algorithm is described in Karney (2012) and was used to generate the statistics reported in this book.

The properties of geodesics which all surveyors should be aware of are:

1. The equator and all meridians are geodesics. They are also the only geodesics on the earth's surface which lie in a plane, and the only ones which form closed loops around the world—the others are endless paths, which never repeat.
2. Geodesics are very similar in shape to great circles, except that they deviate more towards the nearest pole, to take advantage of the flattening of the earth at higher latitudes.
3. For any two points up to 250 km apart on the surface of the WGS84 ellipsoid, the difference between their geodesic separation and their separation along a great circle (assuming the two points lie on a spherical earth of radius 6371 km) is less than one part per million. The maximum differences occur near the equator and near the poles,

where the local curvature reaches its extreme values (see Figure 8.9); elsewhere on the earth's surface, and especially over smaller distances, the difference is always lower than this. If the appropriate spherical radius is calculated at one end of the line using Equation 8.43, then the difference is never more than 0.25 parts per million.

Chapter 9

Map Projections

9.1 THE NEED FOR PROJECTIONS

Chapter 8 has shown how the three-dimensional positions of points on or near the earth's surface can be expressed in terms of their latitude, longitude, and height above an ellipsoid of defined shape. This geodetic co-ordinate system is useful as it gives a precise and straightforward definition of any point's location, in terms of three parameters (east, north and up*) which are convenient to use everywhere around the world.

For many surveying applications it is conventional—and helpful—to represent the positions of points by showing their horizontal positions graphically, and using some other convention to represent their height. Geodetic co-ordinates are shown in this way on 'physical' globes of the earth, which are scale models of an earth-shaped ellipsoid that show terrain of different heights in different colours.

Physical globes with regular ellipsoidal surfaces already incorporate a projection: the peaks of mountains are projected down onto the surface of the ellipsoid, rather than being shown some distance above it. This is a very useful first stage in projection, as it maps all terrestrial detail onto an earth-shaped mathematical surface with very little distortion. However, there are several reasons why this projection by itself is not sufficient for all surveying needs:

1. The ellipsoidal surface is not flat; in fact, it is doubly curved, which means that it cannot be developed (i.e., flattened out, or unwrapped) in the way that is possible with the surface of a cone or a cylinder. This means that the topographic detail on a globe cannot easily be shown on a flat map.
2. There is no fixed or simple relationship between changes in latitude or longitude and movement over the surface of the earth. For instance,

---

* Not exactly 'up' as defined by a plumb-bob, since the ellipsoid is not necessarily exactly parallel to the geoid. The difference in the two directions is called the deviation of the vertical.

an observer on the equator would need to move 1,855 metres to the east to increase his or her longitude by 1 minute; whereas at 40° north (the approximate latitude of New York or Madrid) a movement of just 1,423 metres would have the same effect. Similarly, movements to the north cause increases in latitude which vary depending on the starting position, though the effect is different: movements of 1,843 metres and 1,851 metres respectively would be needed in the example above, to cause an increase of 1 minute in latitude.

3. In surveying, it is often necessary to calculate how far apart two points are when their co-ordinates are known, i.e., the length of the shortest horizontal line between them; and also the angle between two such lines, when they meet or cross at a point.* These calculations are not particularly straightforward to do exactly on the surface of an ellipsoid. The shortest line over a surface representing the earth between two points on that surface is called a geodesic; and because an ellipsoidal surface is based on an ellipse, the shape and length of an ellipsoidal geodesic can only be calculated using elliptic integrals, which do not in general have a closed form (i.e., they do not yield an explicit equation).

One possible solution to Points 2 and 3 above would be to use a sphere rather than an ellipsoid as a mathematical approximation of the geoid. This greatly simplifies the resulting mathematics as geodesics then become great circles, which always lie in a plane and whose lengths are relatively easy to compute. The disadvantage of this solution is that the deviation of the vertical and the geoid-sphere separation both increase considerably, to the point where the mapping of topographic detail onto a spherical surface becomes insufficiently accurate for modern surveying. It was, however, what was routinely done in past times, and much of the early theory of map projection is based on the idea of a spherical earth.

Nowadays, topographic detail is first projected onto an ellipsoid and then either onto a planar surface, or onto a cylinder or a cone—both of which can then be developed (i.e., unwrapped) into a flat surface. The resulting flat surface is then scaled down to form a map.

The remainder of this chapter discusses the main classes of projection used for map making, and the properties which determine their value for surveying purposes.

---

* These two calculations are done repeatedly when performing least-squares adjustment of distance and horizontal angle measurements, as described in Chapter 11.

9.2 USEFUL PROPERTIES OF PROJECTIONS

It is not possible to project a doubly-curved surface onto a planar or developable surface in such a way that the scale of the full-size projection is unity at all places and in all directions. Thus, all such projections involve some degree of distortion on the resulting map, except at certain points or along particular lines. By varying the exact method of projection, it is possible to manipulate the changes in scale so as to reduce or avoid some aspects of distortion on a map, but usually at the expense of causing increased distortion in other respects. In particular, there are three important properties which maps may have, as follows:

1. A *conformal projection* manages the scaling effects such that, at any point on the projection, the scale in all directions is the same.* Such maps are also *orthomorphic*, which means that small shapes on the ellipsoidal surface (e.g., buildings or fields) are shown as the same shape on the map. One result of this is that the angle at which any two horizontal lines cross on the earth's surface is preserved exactly on the map. However, the shortest distance over the ellipsoid between two points on its surface (i.e., a geodesic) does not in general plot as a straight line on the map. In fact, since meridians and the equator are the only geodesics which lie in a plane, they are also the only geodesics which can ever appear as exactly straight lines on any standard projection.

2. An *equal area projection* manages scaling such that, if the scale at a point is unavoidably increased in one direction, then it is correspondingly reduced in the orthogonal direction. Thus, the area of any feature (e.g., a country) is exactly preserved on an equal area map, subject to the quoted scale of the map. However, the shape of the area will not be preserved exactly; and at any point on the map, the scale in one direction (e.g., north–south) will generally differ from the scale in any other direction (e.g., east–west). Small circles drawn on the surface of the earth would thus plot on the projection as ellipses with the same area, but with greater eccentricities in places where the distortion of the projection is higher.

3. In an *equidistant projection*, the scale of the projection is maintained at unity along a particular family of geodesics. This enables distances along those geodesics to be found by simply taking accurate measurements off the map.

---

* On a full-size projection, this scale is called the scale factor. If it is greater than 1, then the size of an object on the projection is greater than its size on the ellipsoid.

No projection can have more than one of these properties over an extended area—and some have none of them, preferring instead to strike a compromise. Conformal projections are generally preferred for surveying purposes, partly because horizontal angles measured in the field (e.g., using total stations) can be plotted directly onto the map.

It is also helpful to think of map projections as having two distinct functions. One is simply to present a region of the world in a form such that the shapes of countries and the distance between features can be estimated to a reasonable accuracy (which will always be limited by the accuracy to which a map can be printed or displayed). The second is to provide a platform for some type of two-dimensional grid system, which can be used to describe the positions of features to very high precision. The amount of distortion which is acceptable in a projection depends on which of these functions it fulfils, and this in turn determines the size of the area that can be mapped. For surveying purposes, the typical requirements are that the scale factor of the grid should not deviate from unity by more than about 1 part in 2000, and that the bearings of geodesics up to 5 km in length should not differ by more than about 5 seconds from that of a straight line plotted on the grid. The guidelines on mapping areas given below are based on these targets.

9.3   COMMON CLASSES OF PROJECTIONS

Aside from the general properties discussed above, there are three important classes of projection, depending on the type of surface onto which the projection is made. These are described below.

9.3.1   Azimuthal Projections

Azimuthal projections were invented (and much used) in classical times, when the shape of the earth was taken to be a sphere. Many of their properties depend on the earth being modelled as a sphere, so the discussion below is presented on that basis.

At any point on the earth's surface, the *azimuthal plane* is defined as the plane passing through the point, and lying normal to the vertical at that point (this is the same as the tangent plane shown in Figure 8.7). An azimuthal projection involves projecting the surrounding region onto that plane; this can be done so as to give a conformal, equal area, or equidistant projection. The point where the plane touches the earth's surface is called the central point (or sometimes the tangent point)—but note that it does not necessarily appear at the centre of the resulting map, which may have been cropped in such a way as to place it off-centre. These projections are called azimuthal because the azimuth (i.e., horizontal) angles between any two straight lines passing through the central point are preserved on

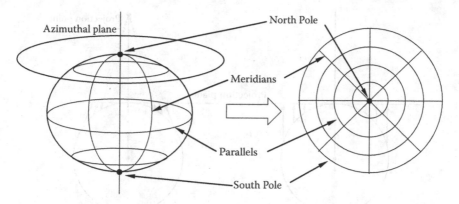

Figure 9.1 Azimuthal projection at the north pole.

the map. Also, all great circles passing through the central point project as straight lines—so any straight line on the map which passes though the central point marks out the shortest path between all the points that it passes through.

An azimuthal projection at the north pole (Figure 9.1) would show the north pole as the central point, with lines of constant latitude (known as parallels) appearing as concentric circles around it. For equidistant and equal area projections, the whole world can be projected, with the south pole appearing as the outermost circle. Meridians would appear as radial lines, crossing the parallels at 90°. If the projection is equidistant, then the parallels would be equally spaced, as in Figure 9.1.

Azimuthal projections can be made at any point on the earth's (spherical) surface, and have distortion which is zero at the central point, and which increases as a function of distance from that point.

As shown in Figure 9.2, the commonest types of azimuthal projection are:

1. *Orthographic,* where each point on the sphere is projected onto the plane along a projector which lies normal to the plane (i.e., the projection point is at infinity). Only half of the earth can be mapped, using one such projection; the earth appears as it would in a photograph taken from a distant point, such as another planet.
2. *Perspective,* where the projectors all pass through a point on the axis normal to the plane which runs through the central point. This gives a view of the earth as it would appear to an astronaut in an orbit round the earth.
3. *Gnomonic* (sometimes call gnomic), a special case of a perspective projection in which the projectors all radiate from the centre of the earth. Since this point lies in the plane of all great circles, the result is that the equator, all meridians, and all shortest paths between any two

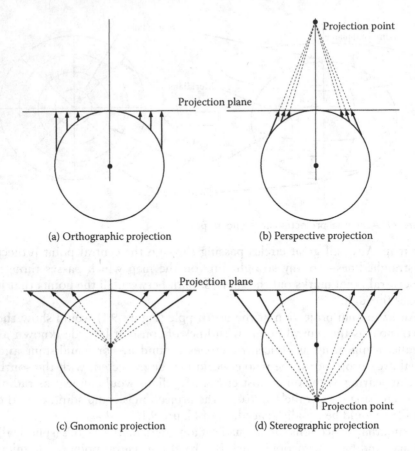

Figure 9.2 Types of azimuthal projection.

points, project as straight lines on the map. Parallels (lines of constant latitude) project as curves similar to those traced out by the shadow of the gnomon on a sundial, which gives the projection its name.

4. *Stereographic,* another special case of a perspective projection, in which the projectors all radiate from the antipode of (i.e., the point exactly opposite) the central point. Some simple trigonometry shows that this gives stereographic projections the property of being conformal.

5. *Equidistant,* this preserves a scale of unity along all the straight lines radiating out from the central point to other points on the map. Equidistant azimuthal projections were particularly useful to bygone emperors who would (of course) use their capital city as the central point, and could then easily see how long it would take to get a message (or perhaps an army) to any part of their empire.

6. *Equal area,* because the scale in the tangential direction increases with distance from the central point, an equal area projection has to reduce the scale in the radial direction by a corresponding amount. This gives rise to quite severe distortions in shape in areas which are distant from the central point.

The value of azimuthal projections to surveyors is limited by the fact that they model the earth as a sphere. Although the earth is very nearly spherical (its diameter at the equator is only about 0.3% greater than the distance between the north and south poles) the level of accuracy needed for surveying is lost if its ellipsoidal shape is modelled as a spherical surface over large areas. For this reason, azimuthal projections of any kind are generally unsuitable for surveying work involving distances greater than about 100 km.*

However, they are often useful for surveying over small areas which are approximately square or circular in shape—and an orthographic azimuth projection is of course what is being used de facto when a simple grid is set up for a construction site, in which the curvature of the earth is ignored altogether.

9.3.2 Cylindrical Projections

The classical cylindrical projection can be visualised as wrapping a sheet of paper round the earth's equator to form a cylinder, and projecting points on the earth's surface out onto it, as shown in Figure 9.3. The distance y is a function of the geodetic latitude, ϕ, of a given point; the function can be defined so as to give an equal area projection, a conformal projection or an equidistant projection in which all meridians are shown with true length.

When the paper is unwrapped, the resulting projection shows parallels as straight horizontal lines, and the meridians as straight, equally-spaced vertical lines. Geodesics running in a north–south direction (i.e., along meridians) are therefore all shown as straight lines. This type of projection has no distortion of shape on the equator and low distortion nearby—so it is particularly suitable for mapping tropical countries.

The transverse form of a cylindrical projection involves wrapping the sheet of paper round a meridian (called the central meridian) rather than the equator, as shown in Figure 9.4. (Actually, this is not strictly a cylindrical projection when used on an ellipsoid, since the meridians are ellipses rather than circles.) The only geodesic which projects as an exactly straight line on a transverse cylindrical projection is the central meridian; geodesics

---

* A larger region (perhaps up to 600 km across) can be mapped if the central point of a conformal azimuthal projection is at or near the north or south pole, as the earth is more nearly spherical in the polar regions.

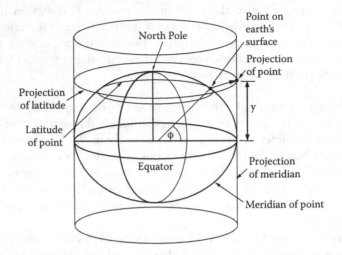

Figure 9.3 Cylindrical projection.

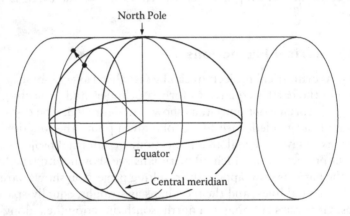

Figure 9.4 Transverse cylindrical projection.

which cross the central meridian at right angles project as lines which are very nearly straight, but not precisely so. Transverse cylindrical projections give maps which have no distortion of shape on the central meridian and low distortion on either side of it, which makes them suitable for countries at any latitude which run in a predominantly north–south direction.

Oblique cylindrical projections also exist—but (like azimuthal projections) they require that the earth is treated as a sphere rather than an ellipsoid, such that the cylinder touches its surface round a great circle other than the equator or a meridian. Such projections are useful for mapping regions which lie along such a line, and work quite well in the polar regions,* where the earth's shape is almost spherical (see Figure 8.9); however, they are less suitable for high-accuracy surveying work over large areas in other parts of the world.

9.3.3 Conical Projections

Conical projections are generally made onto a cone whose axis passes through the north and south poles, and whose surface cuts through the surface of the earth along two parallels, known as the standard parallels. Points on the earth's surface are projected inwards or outwards onto the surface of the cone, along projectors which pass through its axis; the cone's surface is then developed to give a map of the shape shown in Figure 9.5. As in a polar azimuthal projection, parallels project as concentric circular arcs, and the meridians are straight radial lines which cross those arcs at right angles. The spacing of the parallels can be arranged so as to give a projection which is conformal, equal area, or equidistant (with all meridians being shown true length).

The resulting projection has a scale factor of unity on the standard parallels, and no distortion of shape on the parallel midway† between them. A suitable choice of standard parallels can thus give a low-distortion projection of a country running in a predominantly east–west direction.

The angle between neighbouring meridians, and thus the shape of the developed map, depends on the shape of the initial cone, which in turn depends on which two parallels are chosen as the standard parallels. At

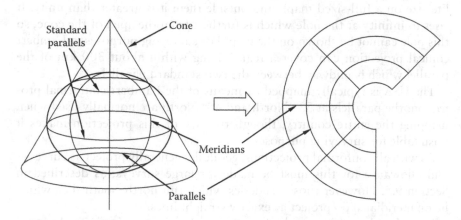

Figure 9.5 Conical projection.

---

* The 'panhandle' of Alaska is generally mapped using an oblique cylindrical projection.
† Strictly, the parallel at the geodetic latitude equal to the cone angle, where the ellipsoidal surface is exactly parallel to the conical surface.

one extreme, the cone becomes a flat circular disc and the result is a polar azimuthal projection, as described above. At the other extreme, the cone becomes a cylinder—either having standard parallels which are equally spaced to the north and south of the equator, or just touching the earth at the equator itself; this, of course, is a cylindrical projection.

9.4 INDIVIDUAL PROJECTIONS

There are over 60 different types of projection in use around the world, usually selected to give the least overall distortion in the context of the size, shape and location of the area which is to be mapped. It is beyond the scope of this book to describe all of these to a useful level of detail; the reader is referred to Bugayevskiy and Snyder (1995) in the first instance. Four projections will, however, be described in further detail, as they are likely to be encountered by all surveyors around the world.

9.4.1 The Lambert Conformal Projection

The Lambert conformal projection is a conical projection, as described in Section 9.3.3—and, as its name suggests, the spacing of the parallels is arranged so as to give maps which are conformal. An ellipsoidal shape and two standard parallels must be specified to define the projection fully. The scale factor of the projection is unity on these two parallels; between them it is less than unity (meaning that projected features would be smaller than life size on a full-sized map), and outside them it is greater than unity. It rises to infinity at the pole which is furthest from the apex of the cone, so this pole cannot be shown on the map. For surveying purposes, a Lambert conical projection can cover a region lying within about 300 km of the parallel which is midway between the two standard parallels.

The USA is typically mapped by means of the Lambert conformal projection: the parallels at 33° North and 45° North are normally used when mapping the entire country, though the size of this projection makes it unsuitable for surveying purposes.

As with all conformal projections, geodesics generally project as curves—and allowance for this must be made for large surveys, as described in Section 9.5. However those geodesics which run north–south (i.e., which lie on meridians) do project as exactly straight lines.

Note that there are several other types of 'Lambert projections'. As noted above, cylindrical and azimuthal projections can be seen as special cases of conical projections, so can be classed as Lambert projections; and Lambert projections are sometimes plotted as equal area projections rather than as conformal ones.

9.4.2 The Mercator Projection

The Mercator projection is a conformal cylindrical projection, with the cylinder touching the earth around its equator. The scale factor of the projection is a function of latitude only: it is unity on the equator, and rises to infinity at the poles. It is therefore impossible for a Mercator projection to show latitudes of greater than about ±80°, and the Mercator projection can only be used for surveying purposes within about 300 km of the equator.

A Mercator projection shows all meridians as straight vertical lines, and all parallels as straight horizontal lines. It thus has the useful property that a straight line drawn between two points on the map cuts each meridian it crosses at the same angle: this is a 'rhumb line', which a navigator could set as a constant compass bearing to travel between the two points. However, rhumb lines are not generally the shortest path between two points, and the corresponding geodesic between the same two points will thus, in general, plot as a curved line on the map. A drawback of this projection is that, above latitudes of about 30°, the scale factor of the map starts to change significantly with increasing latitude. This makes countries near the poles look much larger than they really are by comparison with those near the equator; and also means that, in a country such as Great Britain, the north of Scotland is plotted at a noticeably larger scale than the south of England.

9.4.3 The Transverse Mercator Projection

Also known as the Gauss–Lambert projection, the transverse Mercator (TM) projection is a conformal version of the transverse cylindrical projection described in Section 9.3.2 (see Figure 9.4), and is fully defined by specifying the shape of the ellipsoid, a central meridian, and the value of the scale factor along that meridian (known as the *central scale factor*). The scale factor rises as a function of distance from the central meridian, and the projection is suitable for surveying purposes over an area extending to about 300 km on either side of the chosen meridian.

Great Britain is mapped using a transverse Mercator projection, for the reasons described in Section 9.3. The Airy 1830 ellipsoid is used in classical versions of the mapping, and the 2° West meridian is used as the central meridian because it (more or less) runs up the middle of the country.

If the central scale factor for Britain was set to unity on the central meridian, it would rise to about 1.0008 at the extreme east and west of the country. When making a grid for surveying, it is desirable to have the scale factor as near to unity as possible over the whole of the grid: so the central scale factor for the projection of Britain is actually defined to be about

0.9996* on the central meridian, rising to about 1.0004 at the extreme east and west of the country. This can be visualised as having the cylinder lying slightly inside the central meridian, and cutting through the earth's surface about 180 kilometres on either side of it: features between the two cutting lines are reduced in size when projected onto the cylinder, and those outside them are enlarged.

9.4.4 The Universal Transverse Mercator Projection

The universal transverse Mercator (UTM) projection system provides a low-distortion projection of the whole world, and consists of separate transverse Mercator projections of 60 segments of the earth's surface called *zones*, lying between specified meridians. Each zone covers a 6° band of longitude: Zone 1 runs between 180° W and 174° W, with its central meridian at 177° W, and subsequent zones follow on in an easterly direction. Thus, the central meridian of Zone 29 is 9° W, that of Zone 30 is 3° W, and that of Zone 31 is 3° E. (These are the three zones which are of relevance to the UK.) The central scale factor of each zone is exactly 0.9996, for the reasons discussed above.

The UTM projection is traditionally used in conjunction with the International 1924 ellipsoid, whose dimensions are given in Appendix A. However, it is now also used in conjunction with other ellipsoids (e.g., Clarke 1866 in the USA, and GRS80 or WGS84 elsewhere), so it is important to check which ellipsoid has been used before processing UTM data.

UTM is a useful global projection system, but it is not a universal panacea, because a region which does not wholly lie within 3° of one of the 60 central meridians cannot strictly be plotted on a single UTM zone. This can sometimes be solved by extending a zone slightly; as stated above, TM projections have acceptably low distortion up to 300 km on either side of the central meridian, which would (for instance) allow a variation in longitude of up to 5° from the central meridian, at a latitude of 55°. A further option is to use a nonstandard meridian as the central meridian, as in the British mapping system. The option of butting two or more neighbouring UTM zones together in the region of interest is inappropriate for surveying purposes. As shown in Figure 9.6, the boundaries of neighbouring zones (sometimes called 'gores') are not straight when their underlying cylinders are developed, so can only touch one another at a single latitude. Thus any other point on the common boundary would map to two separate places on the projection, and most lines spanning the boundary would appear discontinuous on the map, where they jump from one zone to the next.

---

* The exact value at the central meridian is 0.9996012717. This (rather improbable) value was chosen to give the best possible fit between the current mapping system in the UK (OSGB36) and its predecessor.

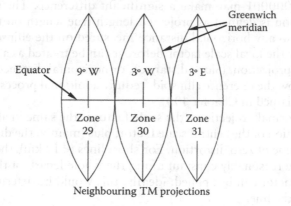

Figure 9.6 Zones of the universal transverse Mercator projection.

9.5 DISTORTIONS IN CONFORMAL PROJECTIONS

As explained at the start of this chapter, the best way of processing topographic data for most surveying purposes is to project it twice—first onto the surface of a suitable ellipsoid, and from there onto a developable surface. Of the various properties which can be retained during the second of these steps, the most useful for surveyors is conformality, i.e., the preservation of small shapes and the angles at which lines on the ellipsoidal surface cross each other. This section therefore concentrates on conformal projections, and discusses what needs to be done to manage the distortions which are (inevitably) present in such projections.

Distortions on a conformal map only become obvious to the naked eye about 45° away from the point or line of zero distortion (which depends on the projection, as explained above). However, some distortion of shape or scale is present over virtually the entire projection, so should always be considered by surveyors.

The two important types of distortion in a conformal projection are (a) that the scale factor varies across the projection, and (b) that the shortest path between two points does not in general plot as a straight line. The standard methods for dealing with both of these are described below.

9.5.1 Scale Factor Distortions

The local scale factor of a projection should be taken into account whenever horizontal distances in the field are converted to distances on the projection. Such distances are routinely measured and recorded to just a few parts per million, so a local scale factor which lies outside the range

0.999999–1.000001 may make a significant difference. The scale factor on a projection is defined as projected length/true length on the full-size projection—so any horizontal distance measured on the ellipsoid must be *multiplied* by the local scale factor before it can be treated as a distance on a conformal projection. (Slope distances, or horizontal distances measured above or below the reference ellipsoid, require additional processing beforehand, as explained in Chapter 10.)

In all conformal projections, the scale factor is the same in all directions and is a function of the central scale factor, plus (mainly) the distance from the point or line of zero distortion. For short lines (< 10 km), the local scale factor will be reasonably constant along the whole length of the line, so a single scale factor can be applied; ideally, this would be determined at the midpoint of the line.

For longer lines, the scale factors should be found at both endpoints, and at the midpoint of the line; a mean scale factor can then be estimated using Simpson's rule, i.e.,

$$S_{mean} = (S_1 + S_2 + 4 \times S_{mid}) / 6 \qquad\qquad 9.1$$

The formula for calculating the scale factor at a point depends on the details of the projection. A precise formula for Lambert conformal projections is given in a document entitled "State Plane Coordinate System of 1983—NOAA Manual NOS NGS 5"[*] and the formula for calculating scale factors in all transverse Mercator projections is given in Appendix D.

In both Lambert conformal and transverse Mercator projections, the local scale factor depends mostly on the distance from the line of zero distortion. A useful approximation (which assumes a spherical earth) is:

$$S = S_0 \left(1 + \frac{d^2}{2 \times R_E^2} \right) \qquad\qquad 9.2$$

where S_0 is the central scale factor, d is the distance of the point from the line of zero distortion, and R_E is the mean radius of the earth (which can be taken as 6.371×10^6 m). For points within 200 km of the line, this formula gives a result which is accurate to within about 2 parts per million; at 500 km from the line, the accuracy is about 12 parts per million.

9.5.2 Distortions of Shortest Paths between Points

As discussed in Chapter 8, the shortest path over the surface of an earth-shaped ellipsoid between two points is called a geodesic. In conformal

---

[*] This is available (May 2013) from http://www.ngs.noaa.gov/PUBS_LIB/ManualNOSNGS5.pdf.

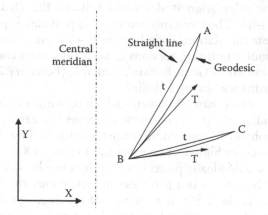

Figure 9.7 The $(t - T)$ correction.

conical projections, the only geodesics which project as straight lines are the meridians; in transverse Mercator (TM) projections, they are the central meridian itself, and (with negligible error*) those geodesics which cross the central meridian at right angles. All other geodesics appear to bulge in the direction of higher scale factor (i.e., away from the central meridian, on a TM projection), when projected. The angles which these lines make with other lines they meet or cross are, however, exactly the same as on the earth's surface.

The effect of this distortion can be seen by reference to Figure 9.7, which is a transverse Mercator plot of three stations, A, B and C. If a total station is set up on station B and stations A and C are observed, the two lines of sight made by the total station will plot as the curved lines on the figure. Because these lines in general 'bulge' by different amounts, the angles they make at B with the corresponding straight lines on the projection (labelled 't') are different—so the angle between the two lines labelled 'T' (i.e., the horizontal angle observed by a total station) will be slightly different from the angle ABC as calculated by trigonometry from the (x,y) co-ordinates of A, B and C.

The angles between the T lines and their corresponding t lines at a point can be calculated (or at least estimated to a higher accuracy than is ever likely to be needed) if the shape of the ellipsoid, the projection parameters, and the positions of both endpoints are known: this calculation is called a '$(t - T)$† correction' or sometimes, an 'arc-to-chord correction'. For the British national grid (and, by extension, for all other transverse Mercator

---

* In the UK, a geodesic of length 600 km which crosses the central meridian at right angles, and is bisected by it, projects as a curved line with a maximum arc-to-chord separation of 16 to 18 mm (depending on its latitude). This is less than 30 parts per billion.
† Pronounced 'tee to tee'.

projections), the calculation is described fully in the Ordnance Survey (1950), pages 14–16. The Ordnance Survey also provides a spreadsheet[*] for many co-ordinate calculations including the $(t - T)$ correction, and plans to include the formulae in future versions of their document entitled 'A guide to coordinate systems in Great Britain', available (February 2013) as a .pdf file from the Ordnance Survey website.[†]

For distances of less than 1 kilometre which lie within 200 kilometres of the central meridian, the $(t - T)$ correction is never more than about 0.5 seconds, so can probably be ignored. Outside these limits, however, it must be taken into account for high-accuracy work. In Figure 9.8, for example, the 7° W meridian should ideally project as a straight line between the latitudes 50° N and 58° N, since it is a geodesic; in fact, it cuts the 54° N latitude at a point which is about 800 metres west of the intercept made by linear interpolation on the projection between the same endpoints. The $(t - T)$ correction for this line is about 12 minutes at each end of the ray.

The $(t - T)$ calculation is tedious to carry out by hand and would normally be done by computer, in the context of adjusting horizontal angle observations. A surveyor should therefore ensure that any adjustment software which works with projected (as opposed to geodetic) co-ordinates takes account of this correction,[‡] before using it for a large-scale survey.

9.6 GRIDS

Having developed (i.e., unwrapped) one of the projections described in Section 9.4 into a plane, it is often convenient to superimpose a right-handed, rectangular Cartesian grid system onto it. This enables the plan position of any point on the map (or ground) to be identified by an (x,y) co-ordinate. The y-axis of the grid is usually defined to lie along a meridian; on a transverse Mercator projection the central meridian is used, and on a Lambert projection the chosen meridian is often called the 'central meridian'. The resulting x and y co-ordinates are then known as Eastings and Northings, respectively, and are usually measured in metres.

The grid system is positioned by specifying the latitude and longitude of a reference point, which is called the *true origin* of the system. However, if the true origin is given the co-ordinates (0,0), it will usually lead to some points having negative co-ordinates, which is undesirable. The true origin may therefore be given a different set of co-ordinates; these are known as "the false co-ordinates of the true origin" (or simply the 'false co-ordinates'),

[*] Available (May 2013) at http://www.ordnancesurvey.co.uk/oswebsite/support/os-net/coordinate-calculations-spreadsheet.html.

[†] Available (May 2013) at http://www.ordnancesurvey.co.uk/oswebsite/docs/support/guide-coordinate-systems-great-britain.pdf.

[‡] LSQ, provided on the book's CRC Press webpage, does so for UTM and British national grid projections.

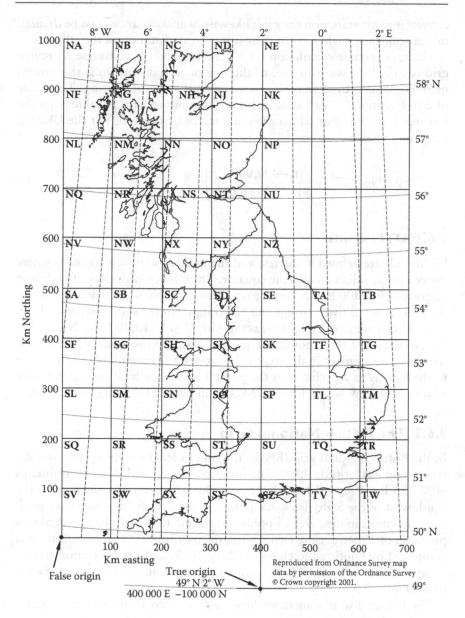

Km Northing

Km easting

False origin

True origin
49° N 2° W
400 000 E −100 000 N

Reproduced from Ordnance Survey map
data by permission of the Ordnance Survey
© Crown copyright 2001.

Figure 9.8 The British national grid.

and the point which thereby acquires the co-ordinates (0,0) is known as the *false origin*.

The grid is attached to the full-size projection, after the central scale factor has been applied. This means that a horizontal distance measured over the surface of the ellipsoid must be *multiplied* by the local scale factor to convert it to a distance on the grid; likewise, a grid distance must be *divided* by the scale factor when converting it to a 'real' distance on the ellipsoid.

The approximate calculation for a scale factor on a transverse Mercator grid is straightforward to use, as the line of zero distortion is the central meridian which, by definition has a constant value of Easting. The value of d in Equation 9.2 can thus be found by subtracting this value from the Easting of the point at which the scale factor is required. For the UK, the equation therefore becomes:

$$S = 0.99960127 \left(1 + \frac{(E - 400,000)^2}{81.179282 \times 10^{12}} \right)$$
9.3

9.6.1 UTM Grids

UTM grids are defined in metres, and the true origin for each zone is where the central meridian crosses the equator. The false co-ordinates of each true origin are (500,000 E, 0 N) for the northern hemisphere, so that all points in the zone have positive Eastings and Northings. For the southern hemisphere, the false co-ordinates are usually set to (500,000 E, 10,000,000 N), so that the Northings of these points are positive also. To distinguish points with the same co-ordinates in different UTM zones, the zone number is attached to the Easting as a prefix: thus the point 1 metre north of the equator with a latitude of 3° W would have UTM co-ordinates of (30,500,000 E, 1 N).

9.6.2 The British National Grid

In the British national grid (BNG), the unit of length is the metre, and the true origin has latitude 49° N and longitude 2° W, with false co-ordinates (400,000 E, –100,000 N). This places the false origin somewhere to the south-west of the Scilly Isles. All points in the British Isles thus have positive grid co-ordinates, and all points on the mainland can be specified to a precision of 1 metre using two six-figure co-ordinates. The specification is completed by quoting the ellipsoid (Airy 1830) and the projection (transverse Mercator with 2° W as the central meridian, and a central scale factor of 0.9996012717).

The British grid is sometimes broken down into 100-kilometre squares with two-letter designators for approximate referencing purposes, as shown in Figure 9.8; thus the co-ordinates of the Cambridge University library's tower (a second-order control point in OSGB36[*]) can be quoted to the nearest centimetre as (544166.76 E, 258409.19 N), or to the nearest kilometre as TL4458.

---

[*] Unfortunately now destroyed by some recent work on the roof.

The formulae to convert between the geodetic (ϕ, λ) co-ordinates of a point and its Eastings and Northings in the British grid are fairly complex, and are unlikely to be needed by most engineering surveyors. If required, they can be found in Ordnance Survey (1950), or in the document entitled 'A guide to coordinate systems in Great Britain'.*

9.6.3 The State Plane Co-ordinate System

For the USA, the size of the country means that using a single projection would cause unacceptable levels of distortion; and most states have decided that the scale factor distortion in each projection should not exceed 100 parts per million. This, plus the wish of each state to have its own self-contained mapping system, means that a total of 107 different projections and associated grid systems (collectively called zones) are used across the main body of the USA, plus a further 16 to cover Alaska, Hawaii, Puerto Rico and the Virgin Islands; collectively, these are known as the state plane co-ordinate system.

Twelve states have just a single zone—either by virtue of their size or shape, or because they chose to accept a slightly larger scale factor distortion. At the other end of the scale, Texas uses five zones, California six, and Alaska ten. Sixty-eight of the projections are Lambert conformal and 54 are transverse Mercator, with one in Alaska being oblique Mercator. Forty-seven states use a single projection method for all their zones, but three do not; Alaska uses all three methods, in different places. The metre is the fundamental unit of length in all states except Arizona—however, much local work is still done in imperial units (the international foot or the U.S. survey foot), using defined conversion factors. The GRS80 ellipsoid is used throughout the system, having supplanted the Clarke spheroid of 1866 which was formerly used for the main body of the USA.

The projection and grid details of all these zones, and the associated methods for calculating local scale factors and $(t - T)$ corrections, are summarised in NOAA Manual NOS NGS 5 entitled 'State Plane Coordinate System of 1983', which can be downloaded (May 2013) from http://www.ngs.noaa.gov/PUBS_LIB/ManualNOSNGS5.pdf.

9.7 BEARINGS ON GRIDS

In surveying, all bearings are measured in a clockwise direction from 'north', as on a compass. However, 'north' can have different meanings,

---

* Available (May 2013) from the web page www.ordnancesurvey.co.uk/oswebsite/support/os-net/coordinate-calculations-spreadsheet.html. This page also provides a downloadable Excel spreadsheet, for performing this and similar calculations.

and so the bearing of a line between two points can have different values, as follows:

1. *True bearings* are measured with respect to the meridian running through a point. True north is where the meridians all meet, on the ellipsoid in use; note that true north on the Airy ellipsoid (for instance) is not in the same physical place as true north on the WGS84 ellipsoid.

2. *Magnetic bearings* are measured with respect to magnetic north, the point on the earth's surface which lies on its magnetic axis. This point is neither stationary, nor coincident with true north on any ellipsoid. The angle between true north and magnetic north is called the *magnetic variation* (or sometimes, the magnetic declination), and can be looked up on some types of map (e.g., aviation maps). If the magnetic variation is west, then magnetic north is to the west of true north, and the variation should therefore be subtracted from the magnetic bearing to obtain a true bearing.* In Great Britain, magnetic variation is currently between about 1° and 4.5° West (depending on location), and reducing by about 6 minutes annually. In the USA, magnetic declination (as it is more commonly known in that country) varies between about 17° East and 19° West, and changes by up to ±12 minutes per year, depending on location.

3. *Compass bearings* (the actual reading from a compass) may differ from magnetic bearings because of nearby ferrous objects, particularly if the compass is mounted in a vehicle. The correction which must be applied to a compass reading to get a magnetic bearing is called the *deviation* of the compass, and is usually plotted as a function of the compass reading. If the deviation is west, then the 'compass north' is to the west of magnetic north, and the deviation should be subtracted from a compass reading to obtain a magnetic bearing.

4. *Grid bearings* are measured with respect to grid north, i.e., the y-axis on the grid system. In the case of a transverse Mercator projection, the y-axis is aligned with the central meridian, so grid north is the same as true north for any point along the central meridian. Elsewhere, the angle between grid north and true north is called the *convergence* of the meridian, and can be calculated as a function of the grid coordinates of the point (Ordnance Survey 1950, 21). Convergence is defined as being positive when true north appears to be to the west

---

* A useful mnemonic is 'variation west, magnetic best [i.e., biggest]—variation east, magnetic least'.

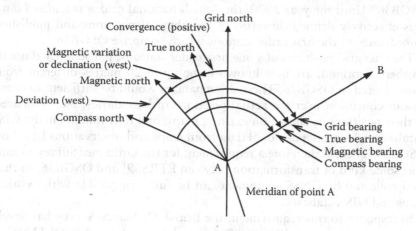

Figure 9.9 Relationship between bearings.

of grid north—on transverse Mercator projections, it is therefore positive at all points to the east of the central meridian. Convergence should be subtracted from a true bearing to obtain a grid bearing—so the grid bearing is smaller than the true bearing when convergence is positive, and larger when it is negative.

The relationship between these directions and angles is summarised in Figure 9.9, for the bearing from point A to point B. Note that the convergence is shown as being positive, and the variation and deviation are both shown as being west.

9.8 THE REALISATION OF THE BRITISH NATIONAL GRID

Section 9.6.2 explained how the BNG is defined in principle; it is useful for surveyors in Britain also to understand how it is realised in practice.

Between 1936 and 1951, about 480 so-called first-order stations were established around mainland Britain, and the geodetic co-ordinates of a few of them were determined as accurately as possible by astronomic observations. Their relative positions were then found by means of triangulation, plus the measurement of a single base line on the Salisbury Plain. The difficulty of measuring distances accurately—and the lack of computing power for adjusting all the observations simultaneously—contributed to inaccuracies in the computed positions of those stations, which are now apparent with the benefit of modern technology. In the meantime, however, all published mapping in Britain has been based on the grid co-ordinates which were originally published for those stations, collectively known as

OSGB36.* Until the year 2000, the British national grid was *realised* (and thus effectively defined) in terms of the physical positions and published co-ordinates of the first-order stations which comprise OSGB36.

The true grid positions of some first-order stations, particularly those in northern Scotland, are now known to be up to 20 metres different from those quoted in OSGB36. These discrepancies would be sufficient to cause serious confusion over the actual positions of boundaries, etc. compared to their positions as shown on existing maps, and to invalidate many GIS† databases which use positional data from maps and observations based on OSGB36. There is therefore a requirement for the Ordnance Survey to support some kind of transformation between ETRS89‡ and OSGB36, so that data collected by GNSS equipment can be fully compatible with existing maps and GIS databases.

In response to this requirement, the British Ordnance Survey has developed a transform called OSTN02™ (supplanting the earlier OSTN97™) which converts ETRS89 Cartesian co-ordinates to grid co-ordinates that are very close to those that would be found§ using OSGB36. The added benefit of this development is that it has allowed the Ordnance Survey to abandon the network of first-order stations as a means of defining the OSGB36 reference frame. Instead, a frame which closely resembles the old OSGB36 is now defined by means of the ETRS89 reference frame, plus the algorithm of OSTN02. OSTN02 is freely available in several forms, including a co-ordinate converter program called GridInQuest,¶ and is now the definitive means for obtaining the OSGB36 grid co-ordinates of any point whose ETRS89 co-ordinates are known. As such, it complements the OSGM02 national geoid model, which has similarly supplanted the national network of benchmarks.

The nature of the errors in the original OSGB36 data means that no simple or homogeneous transformation would provide a satisfactory conversion of co-ordinates between ETRS89 and OSGB36, so a two-stage process is used. Referring to Figure 7.3, the ETRS89 Cartesian co-ordinates are first converted to 'local' geodetic co-ordinates using a GRS80 ellipsoid without any transformation. The transverse Mercator projection and British grid (as defined above) are then applied, to give some initial Eastings and Northings (which are at this stage about 200 metres different from OSGB36 co-ordinates). These are now

---

* In modern parlance, the physical positions and published co-ordinates of these stations is collectively known as a terrestrial realisation frame, or TRF.

† Geographical information system

‡ See Chapter 7, Section 7.6 for a full definition of this terrestrial realisation frame.

§ By measuring angles and distances from nearby OSGB36 control points, and using the published co-ordinates of those control points to determine the co-ordinates of the point in question.

¶ This is downloadable (May 2013) from the Ordnance Survey's OS Net website at http://www.ordnancesurvey.co.uk/oswebsite/support/os-net/grid-inquest.html. GridInQuest also includes the OSGM02 geoid model.

converted to OSGB36 Eastings and Northings using a 'co-ordinate shift model'—a piecewise bilinear interpolation (which can be thought of as a 'rubber sheet' transformation) which firstly takes account of the different shapes and positions of the GRS80 and Airy ellipsoids, but also models the many local distortions present in OSGB36. The accuracy of this conversion process is quoted as 0.4 metres (for 95% of data), though the *relative* accuracy between nearby points can be expected to be somewhat better than that.* The process is therefore quite accurate enough to map GNSS points onto the correct places on an OSGB36 map, but it may cause unacceptable errors in a large engineering project which requires high internal accuracy.

9.9 CO-ORDINATE SYSTEMS FOR ENGINEERING WORKS IN BRITAIN

As explained above, a simple conversion of GNSS data to OSGB36 co-ordinates using OSTN02 will not produce a fully conformal framework of points. In particular, the scale factor might differ from its theoretical value by up to 20 parts per million, and might also depend on the bearing of the line in question. Also, the scale factor changes unpredictably from one region of the country to another, because of the piecewise method originally used to adjust the network of first-order control points.

Many engineering works require a higher precision than this, so an alternative method of establishing a fully conformal grid must be used. Four possible approaches can be summarised as follows:

1. A simple localised co-ordinate system can be used in areas of up to 5 km square, which do not need to be 'tied in' to any larger system. The method for doing this is described in Chapter 3, Section 3.1. Initially, the co-ordinate system would need to be established by conventional means, i.e., using total stations and levels. If necessary, GNSS observations can be incorporated by creating a one-step transform between the local and the ETRS89 co-ordinates of three control points (see Chapter 7, Section 7.6).† The scale factor in the transform should be constrained to be unity.

2. For a larger area, the major control points would typically be surveyed in by GNSS, and the data can be converted directly into UTM

---

* The co-ordinates obtained using GPS and OSTN02 are likely to be only a few cm different from those which might be obtained by resectioning from nearby OSGB36 control points, using their published co-ordinates; however, both sets of co-ordinates may be much further from the 'correct' values which follow from the formal definition of the British national grid, and which were subsequently published by the Ordnance Survey as 'scientific networks' with names such as OS(SN)70 and Ossn80.

† Most GNSS post-processing packages include a facility for doing this.

co-ordinates on the WGS84 or GRS80* ellipsoid. This can usually be done quite easily in the GNSS post-processing software; alternatively, GridInQuest (see above) and the spreadsheet mentioned in Section 9.5.2 both perform this function for ETRS89 data. Subsequent observations might be made by a mixture of GNSS and conventional methods, and a suitable least-squares adjustment program (such as LSQ) will adjust all the observations to high accuracy, using the scale factor calculations and $(t - T)$ corrections described earlier in this chapter. A geoid model would also be required, to convert the ellipsoidal heights recorded by GNSS to the orthometric heights required for the project. This is important, since the local geoid may not be parallel to the WGS84 ellipsoid, so differences in ellipsoidal heights (as measured by GNSS) may differ significantly from differences in orthometric heights (as measured by a level).

3. If the project needs to be tied in with the British national grid, then a fully homogenous co-ordinate system can be set up which will correspond to OSGB36 as closely as possible in the area of interest. First, the OSGB36 Eastings and Northings of three points can simply be picked off a map, and GridInQuest can be used to find their ETRS89 co-ordinates. The points should form an approximately equilateral triangle whose sides should be at least 500 metres long—the places where these points and sides lie will be where the planned transform will fit best with OSGB36.

The orthometric heights of the three OSGB36 points are set to zero, and GridInQuest is used to generate the corresponding ETRS geodetic co-ordinates. A GNSS post-processing package can then be used to generate a best-fit Helmert transform between the two sets of co-ordinates, as described in Chapter 8, Section 8.5.2. Further information will also be needed at this stage: the WGS84 ellipsoid for the ETRS89 geodetic co-ordinates, and the full projection details (as detailed in Section 9.6.2) for the OSGB36 co-ordinates. The OSGB36 points should be specified as having an ellipsoidal height of zero; the transformed Airy ellipsoid will then be fitted to the local geoid, and the ellipsoidal heights from the transform will also be orthometric heights.

The scale factor of the transform should be constrained to be unity, so that the scale factor of the projection gives the correct relationship[†] between grid distances and distances on the ground. The accuracy to which the transformed points will map onto the OSGB36 system will vary depending on the size and location of the project; an indication

---

* The GRS80 ellipsoid has the advantage of being the basis for the OSGM02 geoid model, which facilitates the conversion of ellipsoidal heights to orthometric heights.
† See Section 9.5.1.

will be given by the residual errors which are generated when the transform is computed, which will be reported back by the software.

4. Finally, the Ordnance Survey has published a set of classical transformation parameters to map ETRF89 co-ordinates into OSGB36 (effectively stating the position of the Airy ellipsoid in relation to the ETRS89 ellipsoid). The transform is quoted as having an accuracy of 5 m. The advantage of this transform is that it is fully conformal—in other words, it does not distort the relative positions of points in any way; its drawbacks are (a) that it has a scale factor of 20 parts per million, which would need to be included in all distance computations; and (b), unlike OSTN02 points, these transformed points might have co-ordinates up to 5 m different from those obtained by conventional measurements to nearby OSGB36 control points. (This difference is large enough to make a control point appear to be on the wrong side of a nearby road, when plotted on a map.) The parameters of the transform are given in Appendix A; its main purpose is to provide a simple means for hand-held GNSS devices to show a user's position on the national grid to a level of accuracy which is comparable with that of the device.

Overall, it is probably advisable to use Option 2 for surveying purposes, wherever possible. No transformation is involved, so there is no danger of a 'hidden' scale factor. Also, there is no danger of co-ordinates from different sources, or from slightly different transforms, being mistakenly used in the same calculation or adjustment; UTM co-ordinates are sufficiently different from national grid co-ordinates for their origins to be obvious.

Chapter 10

Reduction of Distance Measurements

The reduction of distance measurements means the process of converting a measured slope distance between two points to the 'reduced' distance over the ellipsoid between the projections of the two points onto the ellipsoid.

It is useful to start by thinking of the simple geometry which would arise if the reference ellipsoid was a flat surface and electromagnetic radiation travelled in a straight line, as shown in Figure 10.1. The relevant quantities which can be measured by a total station are the zenith angle z and the slope distance s. It is clear that we can then write:

$$d = d_R = s \sin z \qquad\qquad 10.1$$

and

$$\Delta h = s \cos z \qquad\qquad 10.2$$

To a first approximation, these quantities will always be correct, even when the curvature of light and of the ellipsoid are considered—but they are not sufficiently accurate to be used directly in surveying calculations.

10.1 CORRECTION FOR THE CURVATURE OF THE ELLIPSOID

The effect of the curvature of the reference ellipsoid is shown (greatly exaggerated) in Figure 10.2. Again, the objective is to reduce the measured distance s to d_R, the reduced distance over the ellipsoid (i.e., the geodesic) between points A_0 and B_0, being the places where lines through points A and B pass normally* through the surface of the ellipsoid.

---

* In other words, the two lines are normal to the piece of the ellipsoid surface that they each pass through.

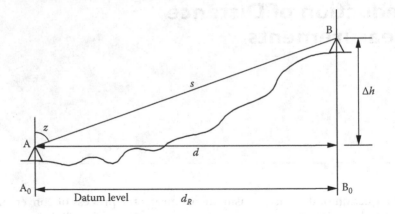

Figure 10.1 Simple calculation of horizontal distance.

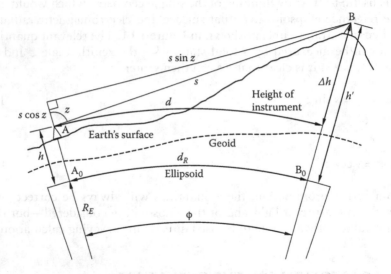

Figure 10.2 Calculation of distances with earth curvature.

The first difficulty which arises is that the lines AA_0 and BB_0 do not lie in a single plane.

If we suppose that the points A, A_0 and B_0 lie exactly in the plane of the paper in Figure 10.2, then the geodetic torsion of the ellipsoid will (in general)* mean that point B, and the lines AB and BB_0 are not quite in the plane of the paper. In addition, the geodesic between points A_0 and B_0 will not generally lie wholly in the plane either.

---

* There is no geodetic torsion along the meridians or around the equator. Geodetic torsion in other places can be calculated using Equation 8.41.

These effects, though, are small. Firstly, the maximum geodetic torsion anywhere on the earth's surface is about 0.11 seconds per kilometre*—so if the distance AB is less than 250 km, then the angle between the line BD and the plane can never be more than 28 seconds. Thus, even if point B has an ellipsoidal height of, say, 2000 metres, it will lie less than 300 mm away from the plane. This means that the true distance AB will differ from its apparent distance (as seen in Figure 10.2) by less than 1 part in 10^{12}. Secondly, as mentioned at the end of Chapter 8, the geodesic distance from A_0 to B_0, and the distance which would be calculated by assuming that both points lie on a spherical earth of radius 6371 km, will not exceed 1 part per million for distances up to 250 km.

It is acceptably accurate, therefore, is to assume that all the lines shown in Figure 10.2 lie in a single plane, and that the path between A_0 and B_0 is a circular arc of radius 6371 km. For any value of s up to 250 km, and for any values of h and h' between 0 and 2000 metres, this basis for calculating d_R results in an error of less than 1 part per million compared to the true geodesic distance between points A_0 and B_0. If the formula given in Equation 8.43 is used to calculate a 'local' value for R_E at one end of the line, then this error drops below 0.25 parts per million, for all distances up to 250 km and for all values of h and h' between –500 and +10,000 metres. Few surveyors are likely to encounter longer distances or more extreme ellipsoidal heights than these.

In Figure 10.2, we can see that:

$$d = (R_E + h)\,\phi \qquad\qquad\qquad 10.3$$

where R_E is the local radius of curvature of the earth, and ϕ is the angle (in radians) subtended by the two points at the centre of the earth. But we can also see that:

$$\tan\phi = \frac{s\sin z}{(R_E + h) + s\cos z} \qquad\qquad\qquad 10.4$$

so we can write:

$$d = (R_E + h)\tan^{-1}\left(\frac{s\sin z}{(R_E + h) + s\cos z}\right) \qquad\qquad\qquad 10.5$$

For all realistic combinations of h, s and z, it turns out that the value of d changes by less than one part per million if h is simply assumed to be zero

---

* This maximum torsion occurs if A and B lie on either side of the equator, with the line between them crossing the equator at 45°; elsewhere on the earth, or for shorter distances, it will be less.

in Equation 10.5. This makes it possible for a total station, or other electro-magnetic distance measurement (EDM) device, to calculate the horizontal distance between two stations to a reasonable degree of accuracy,[*] given a built-in average value for R_E and no information about the height of either station. When the value of h is known, this distance can subsequently be converted to the reduced distance d_R, using the simple formula:

$$d_R = \frac{R_E}{(R_E + h)} d \qquad\qquad 10.6$$

where R_E can be taken to be 6.371×10^6 m. To preserve an accuracy of one part per million, this reduction requires h to be known to an accuracy of $R_E \times 10^{-6}$, or about ± 5 m. Note that h is the height of the geoid above the ellipsoid (see Figure 8.4) plus the height of the observing station above the geoid (i.e., the orthometric height of the station) plus the height of the instrument above the station.

We can also write

$$(R_E + h') \sin \phi = s \sin z \qquad\qquad 10.7$$

i.e.,

$$h' = \frac{s \sin z}{\sin \phi} - R_E \qquad\qquad 10.8$$

Substituting for ϕ using Equation 10.4, and noting that $h' = h + \Delta h$ gives:

$$\Delta h = \frac{s \sin z}{\sin\left(\tan^{-1}\left(\dfrac{s \sin z}{(R_E + h) + s \cos z} \right) \right)} - (R_E + h) \qquad\qquad 10.9$$

As with Equation 10.5, the value of Δh returned by this equation remains virtually unchanged if h is simply assumed to be zero: for distances of up to 20 kilometres and for values of h up to 4000 metres, the resulting error in the value of Δh will be less than one millionth of the measured slope distance.

10.2 CORRECTION FOR LIGHT CURVATURE

A further complication arises, however, from the fact that light (or indeed any other electromagnetic radiation) does not travel in a straight line

---

[*] Subject to an accurate measurement of the angle z, which is more problematic; see Sections 10.2 and 10.4 below.

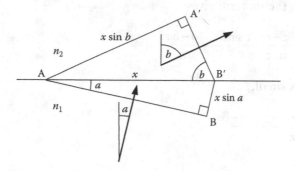

Figure 10.3 Refraction of light (step change of refractive index).

through the earth's atmosphere. This is because the refractive index of the atmosphere reduces with height, making it act like a giant lens. To see the effect of this, first consider what happens when part of a wave front of light travels from a medium with refractive index n_1 to one with lower refractive index n_2, as shown in Figure 10.3. During the time in which the left end of the wave front travels from A to A', the right end (which is still in the denser medium) only travels from B to B'. Thus the wave front changes from making an angle a with the interface to making an angle b with it, where a and b are related by the equation:

$$\frac{x\sin b}{x\sin a} = \frac{\sin b}{\sin a} = \frac{n_1}{n_2}$$

10.10

Now consider light travelling with a zenith angle z from air with refractive index n into air with refractive index $(n + \delta n)$, as shown in Figure 10.4. Using Equation 10.10, we can write:

$$\frac{\sin(z+\delta z)}{\sin z} = \frac{n}{n+\delta n} \approx \frac{n-\delta n}{n}$$

10.11

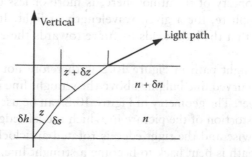

Figure 10.4 Refraction of light (continuous change of refractive index).

Expanding the sin term gives:

$$\frac{\sin z \cos \delta z + \cos z \sin \delta z}{\sin z} = \frac{n - \delta n}{n}$$

10.12

so, since δz is small,

$$\frac{\sin z + \delta z \cos z}{\sin z} = \frac{n - \delta n}{n}$$

10.13

i.e.,

$$\frac{\delta z \cos z}{\sin z} = -\frac{\delta n}{n}$$

10.14

where δz is measured in radians. But $\delta s \cos z = \delta h$, so we can rewrite Equation 10.14 as

$$\frac{\delta h \times \delta z}{\delta s \sin z} = -\frac{\delta n}{n}$$

10.15

whence

$$\frac{dz}{ds} = -\frac{\sin z}{n}\frac{dn}{dh} = \frac{1}{R_P}$$

10.16

where R_P is the radius of curvature of light which is travelling at a zenith angle z through the atmosphere. Conveniently, it turns out that the quantity

$$\frac{1}{n}\frac{dn}{dh},$$

which is a property of the atmosphere, is more or less constant throughout the atmosphere, for a given wavelength of light. It is also negative, which means that the light tends to curve towards the earth, as shown in Figure 10.4.

The actual light path in Figure 10.2 is therefore not a straight line as shown, but a curved line bulging above the straight line joining the instrument and target. The geometry of Figure 10.2 can be used, however, if one imagines a distortion of the picture in which the left edge of the picture is rotated clockwise and the right edge is rotated anticlockwise, so that the curved light path is bent back to become a straight line. The effect of this distortion would be to make the vertical lines through the instrument and

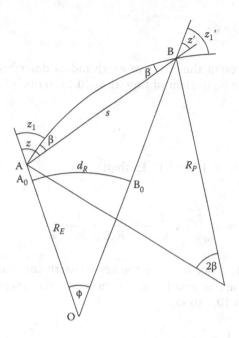

Figure 10.5 Effect of light curvature on zenith angle measurements.

target intersect further down the paper, i.e., to increase R_E. It turns out that the effect of light curvature can be allowed for in exactly this way, by using an 'effective earth radius' which is slightly larger than its actual radius. The mathematical derivation of this effective radius is given below.

Figure 10.5 shows the effect of light curvature on the geometry of Figure 10.2, again greatly exaggerated—the light is assumed to have constant curvature of radius R_P, based on the mean zenith angle of the light path. We can write:

$$\phi = z - z' = \frac{d_R}{R_E} \qquad\qquad 10.17$$

whence

$$\frac{1}{R_E} = \frac{(z - z')}{d_R} \qquad\qquad 10.18$$

It can be seen that, in the absence of light curvature, the change in zenith angle between instrument and target is $(z - z')$. With light curvature, it becomes $(z_1 - z_1')$, where

$$z_1 - z_1' = \phi - 2\beta = \phi - \frac{s}{R_P} \qquad\qquad 10.19$$

Letting R_1 represent the effective earth radius described above, we can therefore write the equivalent of Equation 10.18, namely:

$$\frac{1}{R_1} = \frac{(z_1 - z_1')}{d_R} \qquad\qquad 10.20$$

Using Equations 10.17 and 10.19 then gives:

$$\frac{1}{R_1} = \frac{(z_1 - z_1')}{R_E \phi} = \frac{\phi - \dfrac{s}{R_P}}{R_E \phi} = \frac{1}{R_E} - \frac{s}{R_P R_E \phi} = \frac{1}{R_E} - \frac{s}{R_P d_R} \qquad 10.21$$

Because the difference between the actual earth curvature and the effective earth curvature is small, we can now use the approximate formula given in Equation 10.1 to say:

$$\frac{1}{R_1} = \frac{1}{R_E} - \frac{1}{R_P \sin z} \qquad\qquad 10.22$$

whence, using Equation 10.16,

$$\frac{1}{R_1} = \frac{1}{R_E} + \frac{1}{n}\frac{dn}{dh} = \frac{1}{R_E}\left(1 + \frac{R_E}{n}\frac{dn}{dh}\right) = \frac{1}{R_E}(1 - \kappa) \qquad 10.23$$

where κ is called the refraction constant. Because

$$\frac{1}{n}\frac{dn}{dh}$$

is nearly constant at different heights for any given wavelength, κ is a dimensionless factor which can also be assumed to be constant for a given wavelength of radiation.[*] Typically, κ is quoted as being about 1/7 for visible (or infrared) radiation, and 1/4 for microwave radiation. Using these values for κ, together with a value of 6.37×10^6 m for R_E, gives values for R_1 of 7.43×10^6 m in the case of visible light, and 8.49×10^6 m for microwaves. Other values quoted for the effective earth radius in the literature are 7.52×10^6 m for visible light, and 8.62×10^6 m for microwaves.

---

[*] κ can be defined as being the curvature of light travelling horizontally, as a fraction of the mean curvature of the earth.

We can therefore allow for the effect of light curvature by adapting Equations 10.5 and 10.9 (with h set to zero, as discussed) to give:

$$d = R_1 tan^{-1}\left(\frac{s \sin z_1}{R_1 + s \cos z_1}\right) \qquad 10.24$$

and

$$\Delta h = \frac{s \sin z_1}{\sin\left(tan^{-1}\left(\dfrac{s \sin z_1}{R_1 + s \cos z_1}\right)\right)} - R_1 \qquad 10.25$$

where R_1 is the effective radius of earth curvature for visible light, and z_1 is the actual zenith angle observed by the instrument. The appropriate value for R_1 can be built into an EDM device, enabling it to calculate reasonably accurate horizontal and vertical distances for any sighting within its operating range.

For precise work, however, it is unsatisfactory to simply use the 'horizontal' distance reported by an EDM device, even allowing for the fact that it has not been reduced to the ellipsoid. There are three main reasons for this: firstly, it is not always possible to find out exactly how a given instrument computes its 'horizontal' distances; secondly and most importantly, local atmospheric conditions may mean that κ differs from the values quoted above by up to 100% in either direction,* making the value of R_1 used by the instrument when applying Equations 10.24 and 10.25 quite inappropriate; and thirdly, it is generally better practice to measure an 'uncorrected' slope distance in the field, and compute precise corrections to it subsequently in the office. These corrections will be discussed next.

10.3 CORRECTIONS TO SLOPE DISTANCE MEASUREMENTS

The correction of raw measured distances involves three steps. First, the mean velocity (and therefore wavelength) of electromagnetic radiation along the straight line between instrument and target is not a known, constant value, but depends on the density (largely governed by the temperature and pressure) of the air along the line. For some types of radiation, the humidity

---

* On a 'grazing ray' in particular, when the light path passes close to the surface of the earth, the change in temperature with height can mean that the air nearer the ground is less dense than the air higher up. This causes the light path to bend upwards rather than downwards, giving negative values for κ and values for R_1 which are smaller than R_E.

of the air must also be taken into account. If this is done manually,[*] the correction should be made using a formula or table supplied by the instrument manufacturer, as it is a function of the wavelength and modulation of the radiation used by the instrument. The atmospheric conditions used for this correction are usually the mean values of those measured at each end of the ray; this will be reasonably accurate for pressure, but note that it may be substantially inaccurate for temperature if, say, the instrument and target are on opposite sides of a deep valley.

Secondly, allowance must be made for the fact that the air above the straight line path is slightly less dense than the air on the straight line path, so the radiation will in fact propagate more quickly through it; this, after all, is what causes the light path between the stations to curve as described in Section 10.2. It is therefore necessary to calculate the path through the atmosphere along which the radiation will propagate in the minimum time, and allow for the fact that the mean wavelength along this new path will be longer than the value calculated above. Finally, having now effectively calculated the distance along this new curved path, an 'arc to chord' calculation must be carried out to find the distance along the original straight path.

These last two adjustments are both very small, and indeed tend to cancel each other out. It is therefore common to roll them together into a single correction formula, which makes use of our earlier assumption about the way in which the refractive index (and therefore the propagation velocity) varies with altitude in the earth's atmosphere. The formula quoted by Bomford (1980) is:

$$s' = s\left(1 - \frac{s^2}{24R_E^2}(2\kappa - \kappa^2)\right) \qquad 10.26$$

where s is the distance corrected for mean atmospheric conditions, and κ is the refraction constant defined in Equation 10.23.

In addition to being small, this correction is not particularly sensitive to the variations in κ which can be encountered, as discussed above—so it is generally safe to use the standard value for the electromagnetic radiation used to measure the distance. However, if the EDM uses near-visible (e.g., infrared) radiation, and if the vertical angles at each end of the ray are measured at around the same time as the slope distance, the appropriate value for κ can be found and used in Equation 10.26. The process for doing this is explained in Chapter 12, Section 12.5.

---

[*] Many total stations can do this correction automatically, given the correct temperature and pressure.

10.4 FINAL CALCULATION OF REDUCED DISTANCE

Having made these corrections to s, it might now seem reasonable to calculate d_R using Equation 10.24 followed by Equation 10.6. For really accurate work, however, this is still precluded by unknown effects on the observed vertical angle (z_1) caused by the exact atmospheric conditions at the time of the observation; these effects mean that Equation 10.24 may give an erroneous result, particularly along a steeply sloping ray. Variations in atmospheric conditions quite frequently cause the observed vertical angle to alter by up to 20 seconds*—and an error of this magnitude will cause an error of 8 parts per million in horizontal distance, if the slope between the station is 5°; if the slope angle is 10°, this error rises to 15 parts per million. The effect on vertical distances is even more severe: a 20-second error in the vertical angle will cause an error in Δh of almost 10 cm over a distance of 1 kilometre, or 100 parts per million.

If the height of station A and the height difference of the endpoints (h and Δh respectively, in Figure 10.6) are known to reasonable accuracy,† however, then the value of d_R can be calculated without the need for an observed vertical angle. One way of doing this calculation would be to apply the cosine rule (Appendix A) to the triangle ABO (O being the point

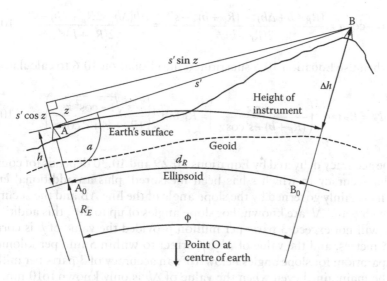

Figure 10.6 Calculation of reduced distance using known heights.

---

* Variations of up to 30 seconds have been noted, but these are less common.
† These values need to be known to within about 5 metres and 10 mm per kilometre of separation, respectively; see the discussion below.

at the centre of the earth where the lines AA_0 and BB_0 meet), to calculate the angle ϕ:

$$\cos\phi = \frac{(R_E+h)^2+(R_E+h+\Delta h)^2-s'^2}{2(R_E+h)(R_E+h+\Delta h)} \qquad 10.27$$

whence

$$d_R = R_E\phi = R_E\cos^{-1}\left(\frac{(R_E+h)^2+(R_E+h+\Delta h)^2-s'^2}{2(R_E+h)(R_E+h+\Delta h)}\right) \qquad 10.28$$

where ϕ is expressed in radians.

However, this method loses accuracy when ϕ becomes small, because the value of $\cos\phi$ has a turning point at $\phi = 0$. If the evaluation of $\cos^{-1}$ is done to (say) 10-figure accuracy, computational inaccuracies will start to exceed one part per million when the rate of change of $\cos\phi$ with respect to ϕ (i.e., $\sin\phi$) falls below 1×10^{-4}. This happens when ϕ is less than 0.0001 radians, i.e., when d_R is less than about 600 metres.

A more reliable approach, which works for all measurable distances, is to use the cosine rule to calculate the value of $\cos z$, in Figure 10.6:

$$\cos z = -\cos a = \frac{(R_E+h+\Delta h)^2-(R_E+h)^2-s'^2}{2(R_E+h)s'} = \frac{\Delta h(\Delta h+2R_E+2h)-s'^2}{2(R_E+h)s'} \qquad 10.29$$

And then use Equation 10.5 substituted into Equation 10.6 to calculate d_R:

$$d_R = R_E\tan^{-1}\left(\frac{s'\sin z}{(R_E+h)+s'\cos z}\right) = R_E\tan^{-1}\left(\frac{s'\sqrt{1-\cos^2 z}}{(R_E+h)+s'\cos z}\right) \qquad 10.30$$

The accuracy delivered by Equations 10.29 and 10.30 depends (of course) on the accuracy to which s has been measured, plus an additional error which is mainly governed by the slope angle of the line AB and the accuracy to which h and Δh are known. For slope angles of up to 20°, this additional error will not exceed 3 parts per million provided the value of h is correct to ±5 metres, and the value of Δh is correct to within 5 mm per kilometre of separation; for slope angles below 10°, an accuracy of 3 parts per million will be maintained even when the value of Δh is only known to 10 mm per kilometre of separation. These levels of error remain fairly constant for all distances between 10 m and 250 km, and reduce to 2 parts per million if h is known to within ±0.5 metres.

Depending on the distance between A and B, the required accuracy for Δh can be achieved by levelling, by GNSS, or by trigonometric heighting as described in Chapter 12. Note, however, that the value of Δh used in Equation 10.29 involves the heights of the tripods at each end of the measured distance, as well as the height difference between the two stations (see Figure 10.6).

10.5 SLOPE DISTANCES

An alternative approach is to use the station-to-station slope distance rather than d_R, when compiling data for an adjustment calculation. If this route is taken, then it may involve one further operation on the distance computed in Equation 10.26; since this is a distance from instrument to target, it will probably need to be adjusted manually to give a slope distance between the two actual stations before it can be fed into a network adjustment program. An approximate formula for this correction can be derived with reference to Figure 10.7, in which A and B represent the instrument and target, and E and F the stations over which they are positioned.

We start by rotating the line AB about X, the point where it crosses the angular bisector of OA and OB (where O is the centre of the earth), until it is parallel with the line between the two stations. This gives the line $A'B'$,

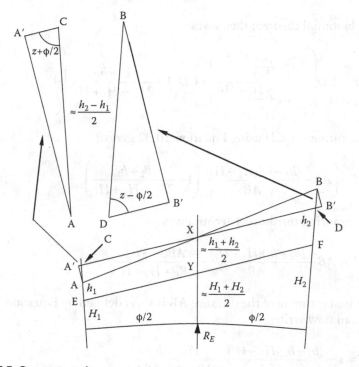

Figure 10.7 Correction of measured slope distance.

which is then trimmed down to give the line CD. Because ϕ is small, the lengths AC and BD are both approximately $(h_2 - h_1)/2$, so we can write:

$$CD \approx A'B' - \frac{h_2 - h_1}{2} \left\{ \cos\left(z + \frac{\phi}{2}\right) + \cos\left(z - \frac{\phi}{2}\right) \right\}$$

$$\text{giving } CD \approx A'B' - (h_2 - h_1)\left(\cos z \cos\frac{\phi}{2}\right) \qquad 10.31$$

i.e., recognising that ϕ is very small and using Equation 10.2,

$$CD \approx AB - (h_2 - h_1)\cos z \approx AB - (h_2 - h_1)\frac{H_2 - H_1}{AB} \qquad 10.32$$

The triangles OCD and OEF are similar, so we can now use the approximate distances OY and OX shown in the figure, and write:

$$EF = CD \times \frac{OY}{OX} = CD \times \frac{R_E + \dfrac{H_1 + H_2}{2}}{R_E + \dfrac{H_1 + H_2}{2} + \dfrac{h_1 + h_2}{2}} \qquad 10.33$$

The binomial theorem then gives

$$EF \approx CD\left(1 - \frac{\dfrac{h_1 + h_2}{2}}{R_E + \dfrac{H_1 + H_2}{2}}\right) = CD\left(1 - \frac{h_1 + h_2}{2R_E + H_1 + H_2}\right) \qquad 10.34$$

Substituting for CD using Equation 10.32 gives

$$EF = \left(AB - \frac{(h_2 - h_1)(H_2 - H_1)}{AB}\right)\left(1 - \frac{h_2 + h_1}{2R_E + H_2 + H_1}\right) \qquad 10.35$$

which, ignoring third-order terms, gives

$$EF \approx AB - \frac{(h_2 - h_1)(H_2 - H_1)}{AB} - \frac{AB(h_2 + h_1)}{2R_E + H_2 + H_1} \qquad 10.36$$

Our best estimate of the distance AB is s' (as defined by Equation 10.26), so we can now write

$$s'' = s' - \frac{(h_2 - h_1)(H_2 - H_1)}{s'} - \frac{s'(h_2 + h_1)}{2R_E + H_2 + H_1} \qquad 10.37$$

where H_1 and H_2 are the ellipsoidal heights of the two stations, and h_1 and h_2 are the heights of the instrument and target above their respective stations. For normal distance measurements, the value of $(h_2 + h_1)$ is so small in comparison to R_E that we can ignore H_1 and H_2 in the second term of Equation 10.37 and write

$$s'' = s' - \frac{(h_2 - h_1)(H_2 - H_1)}{s'} - \frac{s'(h_2 + h_1)}{2R_E} \qquad 10.38$$

This means that we do not need to know the absolute values of H_1 and H_2 to find s'', but merely the difference between them—which typically only needs to be known to within about 0.1 m to preserve an accuracy of 1 part per million* in Equation 10.38. The values of h_1 and h_2 need to be known to within about 2 mm, which can easily be achieved by measuring the heights of instrument and target with a tape. A suitable value for $(H_2 - H_1)$ would nowadays typically be found using GNSS; where this is not practicable, it can be found by levelling, or by using trigonometric heighting as described in Chapter 12.

The station-to-station slope distance can also be used to calculate the reduced distance, d_R: simply apply Equations 10.29 and 10.30 using s'' in place of s', and taking Δh to be the station-to-station height difference, rather than the instrument-to-target value.

A worksheet for adjusting measured slope distances using the formulae developed in Sections 10.3 and 10.5 is given in Appendix H.

10.6 SUMMARY

The essential points covered by this chapter are as follows:

1. Total stations are capable of calculating horizontal distances, but only at the altitude of the observing station. To obtain a reduced horizontal distance on the ellipsoid, Equation 10.6 must be applied.
2. All observed vertical angles are affected by atmospheric effects. It is possible to make some allowance for this by using an effective earth radius in place of the real one, but this makes assumptions about the atmosphere which may not be true at the time of observation. Under these circumstances, the calculation of a vertical distance from an

---

* The suggestions for accuracy given throughout this chapter are only guidelines for measurements taken in normal terrain. A surveyor who is in any doubt about the effect that approximate data might have on a formula should compute the formula twice, using extreme values for the data, and see whether the change is significant.

observed slope distance and vertical angle will be in error—as will the horizontal distance, if the ray has a slope of more than a few degrees.

3. The remedy is to calculate the reduced distance or the station-to-station slope distance using the height difference between the two stations, which must therefore be known. Although the atmospheric effects mentioned above also affect the measurement of slope distance itself, the effect is very small in practice.

4. A measured and corrected slope distance (instrument to target) can be directly converted to a reduced horizontal distance between the stations, if the height of one station is known to better than 5 metres. It can also be converted to a station-to-station slope distance, if this data can be used by an adjustment program.

Chapter 11

Adjustment of Observations

11.1 INTRODUCTION

The term 'adjustment' suggests that some form of cheating might take place during the process of converting surveying observations into results. This is not necessarily the case. As explained in Chapter 2, surveying observations always have two particular properties:

1. They contain errors (random, systematic, and the occasional gross error).
2. The system of observations should always be such that more observations have been taken than would be strictly necessary to obtain the required result (i.e., there is redundancy in the system).

This combination of properties means that no set of surveying observations is ever exactly consistent. If, for instance, the purpose of the observations is to fix the position of a new station, there will be no position which will exactly concur with all the data. All we can do is to choose a position for the point which gives the best agreement with the observational data that we have, and say that this is the most likely position of the point. The process of finding this 'best' position is called adjustment. It only involves cheating if, in computing a 'best' position for the point, we ignore some of our observations for the sole reason that they do not appear to agree well with the other observations.

The adjustment of large quantities of observations involving several stations whose positions are unknown is a tedious job, and is best done by computer. Several programs exist for this purpose, and are able to take detailed account of what constitutes the 'best' fit of the available data. The process they use is called *least-squares adjustment*, and will be described later in this chapter. Often, though, it is adequate to use a simpler method which does not need elaborate computer software; one such method is called a Bowditch adjustment, and a simplified version of this adjustment is described in the next section.

11.2 THE BOWDITCH ADJUSTMENT

This adjustment method is best suited to a traverse (either in a horizontal plane or in the vertical direction). A basic version of the general method will be described, with reference to a simple example. All distances will be assumed to be small, so that the $(t - T)$ corrections described in Chapter 9 can be ignored.

A classic traverse to find the grid* positions of two new stations is shown in Figure 11.1. The grid positions of stations A, B, E, and F are already known, and the positions of stations C and D need to be found.

The typical scheme of observations to find these positions with some degree of redundancy is also shown on the figure. Angles are measured as shown, at stations B, C, D and E; and the horizontal distances BC, CD and DE are also measured. Note that the angles are measured using the previous station as the reference object: at D, for instance, C is used as the reference object, giving an observed angle CDE greater than 180°.

Given these data, it is first possible to calculate the grid bearing of the line BA by simple trigonometry—then, by simply adding the measured angle ABC, the grid bearing of the line BC. Knowing this bearing, and the length of the line BC,[†] we can calculate a preliminary grid position for C.

We can now repeat this process, to calculate preliminary grid positions for D, and then for E. If all the readings were totally error-free, the preliminary grid position we would calculate for E would be exactly the same as its known grid position, which we already have as part of our data.

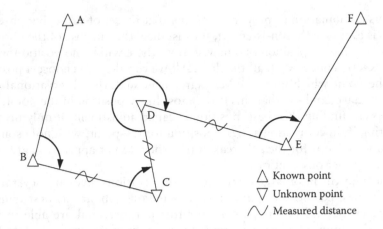

△ Known point
▽ Unknown point
◠◡ Measured distance

Figure 11.1 Four-point traverse.

---

* As defined in Chapter 9, Section 9.6
† The measured distance must first be reduced to an ellipsoidal distance (Chapter 10) and then multiplied by the local scale factor (Chapter 9) to convert it to a grid distance.

Inevitably, though, this will not be the case, and the fact that E has not turned out to be exactly where it is known to be means that the calculated positions for C and D are probably not correct either.

In its simplest form, the Bowditch adjustment assumes that the errors in observation, which have given rise to the misplacement of E, are uniformly distributed throughout the traverse: so that C (which is 1/3 of the way through the traverse) should be adjusted by 1/3 of the amount by which E must be adjusted, and D (which is 2/3 of the way through the traverse) should be adjusted by 2/3 of the same amount. Suppose, for instance, that the calculated position for E turns out to be 18 mm north and 12 mm west of the correct position: the best guess for the actual position of C is to subtract 6 mm from the Northing and add 4 mm to the Easting of the preliminary position. Likewise for D, we subtract 12 mm from the Northing, and add 8 mm to the Easting.

More elaborate forms of the adjustment also exist. Suppose, for instance, that the distance BC is 1/5 of the total distance BC + CD + DE. A better adjustment might be to adjust C by 1/5 of the adjustment needed for E, rather than by 1/3 as described above. In general, however, the extra work involved in these more elaborate adjustments does not result in a noticeably better set of co-ordinates for the unknown points.

A calculation sheet for the application of a simple Bowditch adjustment to a two-point traverse of the type described above is shown in Figure 11.2. The calculation starts by filling in all the shaded squares, which are either data or observations.

The next stage is to calculate all the bearings down the right-hand side of the sheet. The bearing of the line from B to A is calculated first, from the given grid co-ordinates. Adding the angle ABC* to this value, and subtracting 360° if necessary, gives the bearing from B to C. Adding or subtracting 180° now gives the bearing from C to B, and the process can be repeated until a computed value of the bearing from E to F is obtained. This is then compared with the value which is computed from the known co-ordinates of E and F.

If the agreement between these bearings is reasonable (i.e., not more than about twice the error that might be expected from a single angle measurement[†]), the main calculation can now proceed. The next step is to convert all the observed horizontal distances to grid distances, filling in the spaces provided. This typically involves two operations:

---

* Defined as the angle measured at B, using A as a reference object and swinging round to C.
† If n measurements with standard deviation σ are added together, the standard deviation of the result is $\sigma\sqrt{n}$. Since there are four angle measurements in this traverse, we might expect the final bearing to be in error by twice that of an individual measurement.

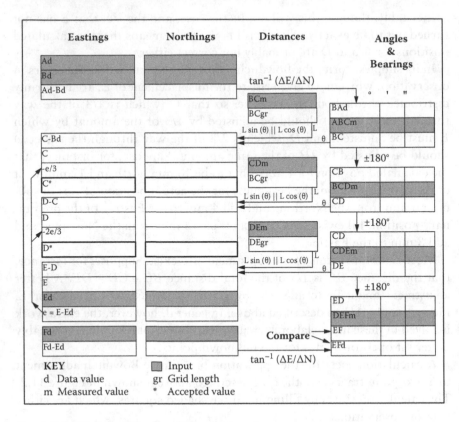

Figure II.2 Worksheet for Bowditch adjustment.

1. Multiplying the measured horizontal distance by the factor

$$\frac{R_E}{R_E + h},$$

   where R_E is the radius of the earth (6.371×10^6 m), and h is the height of the instrument above the ellipsoid (see Chapter 10, up to Equation 10.6). Note that the ellipsoidal height of the instrument includes the height of the instrument above the geoid (i.e., its orthometric height), plus the height of the geoid above the ellipsoid in the area where the observation is made. Note also that this adjustment changes the distance by less than 5 parts per million if h is less than about ±25 m.
2. Multiplying the distance calculated above by the local scale factor, as described in Chapter 9, Section 9.5.1. For all but the largest traverses, this factor will be the same for all the distances in the traverse.

Having done this, the Easting and Northing components of the vector BC can be calculated, using the computed bearing and distance from B to

C. Adding the grid co-ordinates of B to this vector gives the preliminary co-ordinates for C. This process is then repeated to find preliminary co-ordinates for D and E as well.

Finally, the co-ordinates which have just been computed for E are compared with its known co-ordinates, and the error or misclosure is found. As described above, 1/3 of this misclosure is now subtracted from C, and 2/3 from D, to yield accepted values (i.e., best guesses) for the co-ordinates of these two points.

A sample calculation of a Bowditch adjustment is given in Appendix E.

11.3 LEAST-SQUARES ADJUSTMENT

The goal of adjustment is to choose the most likely co-ordinates for points whose positions are unknown, given the available readings. Once these co-ordinates have been chosen, the calculated angles and distances between these points and the fixed points will not all be exactly the same as those which were observed; any difference between each calculated and observed value must be presumed to be the error in the reading. Assuming that these errors have a statistically normal distribution, it can be shown that the most likely co-ordinates for the unknown points are those which yield the smallest value when the squares of all the errors are added together. The process of finding these co-ordinates is therefore known as *least-squares adjustment*.

The exact process involved in least-squares adjustment[*] is best explained by reference to a simple example. Suppose we wish to find the most likely Easting and Northing co-ordinates of the unknown point C in Figure 11.3, and have used two known points, A and B to take three measurements: (1)

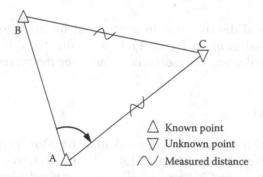

△ Known point
▽ Unknown point
◠◡ Measured distance

Figure 11.3 Sample problem for least-squares adjustment.

---

[*] In fact, there are two distinct methods for carrying out least-squares adjustment; this one is called 'variation of co-ordinates'.

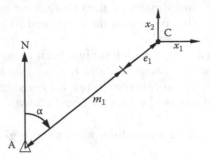

Figure 11.4 Error in a distance measurement.

the distance AC, (2) the distance BC, and (3) the angle BAC. Because of errors in these measurements, it will not in general* be possible to find any position for C which will exactly concur with all the data.

The process starts by taking a guessed position for C, and then seeing how this guess should be altered, in the hope of bringing the sum of the squares of the errors to a minimum. Taking the (known) co-ordinates of A and the (guessed) co-ordinates of C to be (A_N, A_E) and (C_N, C_E) respectively, we can calculate the distance and bearing from A to this position for C (Figure 11.4):

$$d_{AC} = \sqrt{(C_N - A_N)^2 + (C_E - A_E)^2} \qquad\qquad 11.1$$

$$\alpha = \tan^{-1}\left(\frac{C_E - A_E}{C_N - A_N}\right) \qquad\qquad 11.2$$

This calculated distance will in general not be equal to the measured distance, labelled as m_1 in Figure 11.4. We define the difference between the calculated value and the measured value[†] to be the current error for this reading, i.e.,

$$e_1 = d_{AC} - m_1 \qquad\qquad 11.3$$

Now consider the two degrees of freedom we have for altering the position of C, denoted as x_1 and x_2 in Figure 11.4. In Figure 11.5, we can see that adding the quantity x_1 to C's Easting will increase the calculated distance AC

---

* It is always possible that the errors in a set of observations will cancel each other out, to give the impression that they do not exist at all. Adding further redundancy to a set of observations reduces the likelihood of this undesirable possibility.

† The actual distance measured must first be reduced to the ellipsoid (see Chapter 10), and must then be multiplied by the local grid scale factor before being used in this equation.

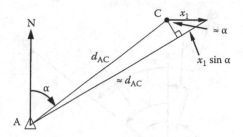

Figure 11.5 Change in distance when C moves east.

by approximately ($x_1 \sin \alpha$), provided x_1 is small compared to d_{AC}. Likewise in Figure 11.6, we can see that adding the quantity x_2 to C's Northing will increase AC by approximately ($x_2 \cos \alpha$), again provided x_2 is small.

We can therefore expand Equation 11.3 to include the effects of x_1 and x_2, as follows:

$$e_1 = d_{AC} + x_1 \sin\alpha + x_2 \cos\alpha - m_1 \qquad\qquad 11.4$$

Likewise, for the second measurement, we can write:

$$e_2 = d_{BC} + x_1 \sin\beta + x_2 \cos\beta - m_2 \qquad\qquad 11.5$$

where β is the calculated bearing from B to the current position of C, and m_2 is the measured distance from B to C.

The third measurement is an angle, and so must be treated slightly differently. Looking at Figure 11.7, we can first write the equivalent of Equation 11.3:

$$e_3 = (\alpha - \gamma + 2\pi) - m_3 \qquad\qquad 11.6$$

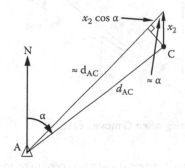

Figure 11.6 Change in distance when C moves north.

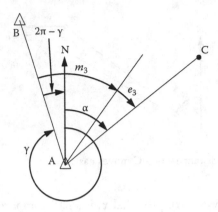

Figure 11.7 Error in an angle measurement.

where γ is the calculated bearing* from A to B, based on their (known) co-ordinates.

Now in Figure 11.8 we can see that adding x_1 to C's Easting will increase the calculated bearing from A to C by approximately

$$\frac{x_1 \cos \alpha}{d_{AC}}$$

provided x_1 is small compared to d_{AC}, while Figure 11.9 shows that adding x_2 to C's Northing will reduce the calculated bearing by approximately

$$\frac{x_2 \sin \alpha}{d_{AC}} .$$

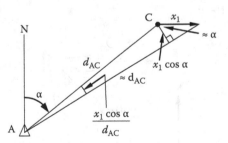

Figure 11.8 Change in bearing when C moves east.

---

* If the distances are long, the bearings α and γ first need to be adjusted using the $(t - T)$ correction, as described in Chapter 9, Section 9.5.2.

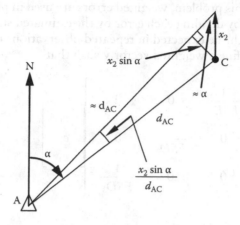

Figure 11.9 Change in bearing when C moves north.

We can therefore write the equivalent of Equation 11.4 for angular measurements, namely:

$$e_3 = (\alpha + \gamma - 2\pi) + \frac{x_1 \cos\alpha}{d_{AC}} - \frac{x_2 \sin\alpha}{d_{AC}} - m_3 \qquad 11.7$$

Equations 11.4, 11.5 and 11.7 can now be written in matrix form, giving:

$$\begin{bmatrix} e_1 \\ e_2 \\ e_3 \end{bmatrix} = \begin{bmatrix} \sin\alpha & \cos\alpha \\ \sin\beta & \cos\beta \\ \dfrac{\cos\alpha}{d_{AC}} & -\dfrac{\sin\alpha}{d_{AC}} \end{bmatrix} \begin{bmatrix} x_1 \\ x_2 \end{bmatrix} - \begin{bmatrix} m_1 - d_{AC} \\ m_2 - d_{BC} \\ m_3 - (\alpha + \gamma - 2\pi) \end{bmatrix} \qquad 11.8$$

or

$$\mathbf{e} = \mathbf{A}\,\mathbf{x} - \mathbf{k} \qquad 11.9$$

which shows that the three-dimensional vector **e** is a linear function of the two-dimensional vector **x** (correct provided the elements in **x** are small), with **A** and **k** containing only constants. Clearly the goal is to set the elements in **x** so as to make the components in **e** as small as possible.

However, there is one further complication to be overcome first: whereas e_1 and e_2 are distances, e_3 is an angle, so its value cannot be compared directly with the other two values. Even in the case of the two distance measurements, it may be that one measurement is known to be much more accurate than the other, so that, say, a 1 mm error in measurement 1 is as likely as a 10 mm error in measurement 2.

To overcome this problem, weighted errors are used in place of the actual errors, obtained by dividing each error by the estimated standard deviation (ESD) which might be expected in repeated observations of the same measurement. We define a weighted vector **v** such that

$$
\begin{bmatrix} v_1 \\ v_2 \\ v_3 \end{bmatrix} = \begin{bmatrix} \dfrac{1}{ESD_1} & 0 & 0 \\ 0 & \dfrac{1}{ESD_2} & 0 \\ 0 & 0 & \dfrac{1}{ESD_3} \end{bmatrix} \begin{bmatrix} e_1 \\ e_2 \\ e_3 \end{bmatrix} \qquad 11.10
$$

or

$$\mathbf{v} = \mathbf{R}\,\mathbf{e} \qquad\qquad 11.11$$

where **R** contains the reciprocal of the ESD for each reading along its diagonal. Substituting for e using Equation 11.9 gives:

$$\mathbf{v} = \mathbf{R}\,\mathbf{A}\,\mathbf{x} - \mathbf{R}\,\mathbf{k} \qquad\qquad 11.12$$

and now the goal is to find the values for **x** which give the smallest result when the scalar elements of **v** are squared, and added together (i.e., the dot product of **v** with itself).

It is clear at this stage that Equation 11.12 can be used for problems involving more than three observations or two variables, as in our simple example. For bigger problems, **v**, **R**, **A** and **k** have a row for each measurement, with the elements in **A** and **k** being calculated from the initial guessed geometry, as shown in the example. The remainder of this explanation will therefore assume that there are m measurements, and n variables—with m being larger than n, to reflect the redundancy in the system of measurements.

Consider now how the ith element in **v** is obtained from Equation 11.12. We can write:

$$v_i = r_{ii}\,a_{i1}\,x_1 + r_{ii}\,a_{i2}\,x_2 + \ldots + r_{ii}\,a_{in}\,x_n - r_{ii}\,k_i \qquad 11.13$$

where the first subscript refers to the row of the matrix, and the second to the column.

Now consider the partial differential of v_i with respect to one of the variables, x_j say. If all the other elements in **x** are held constant, this is:

$$\frac{\partial v_i}{\partial x_j} = r_{ii} a_{ij} \qquad\qquad 11.14$$

The partial differential of the square of v_i with respect to x_j is therefore:

$$\frac{\partial(v_i^2)}{\partial x_j} = 2\, v_i \frac{\partial v_i}{\partial x_j} = 2\, v_i\, r_{ii}\, a_{ij} \qquad 11.15$$

The partial differential of the sum of the squares of *all* the elements in v with respect to x_j is equal to the sum of the individual partial differentials,* i.e.,

$$\frac{\partial\left(\sum\limits_{i=1}^{i=m} v_i^2\right)}{\partial x_j} = \sum_{i=1}^{i=m} \frac{\partial(v_i^2)}{\partial x_j} = 2(v_1\, r_{11}\, a_{1j} + v_2\, r_{22}\, a_{2j} + \ldots + v_m\, r_{mm}\, a_{mj}) \qquad 11.16$$

The sum of the squares of the elements in v can be thought of as a scalar field, which is a function of the n-dimensional space defined by x. At the minimum,† the partial differential of this scalar field will be zero with respect to *each* variable in x; in other words, the expression on the right-hand side of Equation 11.16 must be equal to zero for *all* values of j between 1 and n. Expressing this requirement in matrix form gives:

$$2\begin{bmatrix} a_{11} & a_{21} & \ldots & a_{m1} \\ a_{12} & a_{22} & \ldots & a_{m2} \\ \vdots & \vdots & \vdots & \vdots \\ a_{1n} & a_{2n} & \ldots & a_{mn} \end{bmatrix} \begin{bmatrix} r_{11} & 0 & \ldots & 0 \\ 0 & r_{22} & \ldots & 0 \\ \vdots & \vdots & \vdots & \vdots \\ 0 & 0 & \ldots & r_{mm} \end{bmatrix} \begin{bmatrix} v_1 \\ v_2 \\ \vdots \\ v_m \end{bmatrix} = \begin{bmatrix} 0 \\ 0 \\ \vdots \\ 0 \end{bmatrix} \qquad 11.17$$

Noting that the left-most matrix is in fact the transpose of **A**, dividing both sides by 2, and using Equation 11.12 to substitute for v, we can therefore write:

$$A^T R v = A^T R R A x - A^T R R k = 0 \qquad 11.18$$

or

$$A^T W A x = A^T W k \qquad 11.19$$

where $W = (R \times R)$, and is called the weight matrix. This then gives a solution for x, namely:

---

* This is quite a difficult sentence to understand properly, but it is worth the effort!
† Another way of stating the (first-order) conditions for the minimum of a scalar field is that the grad vector should be zero, which is what Equation 11.17 expresses.

$$\mathbf{x} = (\mathbf{A}^T \mathbf{W} \mathbf{A})^{-1} - 1 \mathbf{A}^T \mathbf{W} \mathbf{k} \qquad\qquad 11.20$$

Provided the errors vary linearly with the elements in $\mathbf{x}$, these adjustments, when applied to the current positions of all the unknown points, will yield the lowest possible sum of the squares of the errors. In practice, of course, the variations in the errors are not exactly linear with respect to the adjustments, even when the adjustments are quite small, but provided they are reasonably linear, the new positions for the unknown points will be much closer to the optimum positions than the old ones. The next stage therefore is to adjust the positions of the unknown points as indicated, recompute the values in $\mathbf{A}$ and $\mathbf{k}$, and re-apply Equation 11.20. This process is then repeated until the required adjustments become negligible, i.e., until $\mathbf{x} \approx 0$. Essentially this is a form of Newton's method for finding the roots of an expression—and just as in Newton's method, it is essential that the initial guesses for the unknown quantities are reasonably close to the correct values, for the method to converge.

11.4 ERROR ELLIPSES

An extremely useful by-product of the least-squares adjustment process is an indication of the likely accuracy to which the unknown points have been found, based on the geometry of the observations plus the apparent distribution of errors at the end of the adjustment.

The first stage is to see whether the initial estimates for the standard deviations on the readings appear to be valid, now that the residual weighted errors have been reduced to their smallest possible size. If they were valid, then the standard deviation of the values in the weighted error vector $\mathbf{v}$ should be close to unity.

Statistically, if m independent samples (v_1 to v_m) are drawn from a population whose average value is already known, the standard deviation of the population can be estimated by the equation:

$$\sigma_v^2 = \frac{1}{m} \sum_{i=1}^{m} (v_i - \bar{v})^2 \qquad\qquad 11.21$$

where σ_v is the standard deviation, v_i is one of the samples, and $\bar{v}$ is the known average.

In the case of the samples contained in the weighted error vector $\mathbf{v}$, the average value is in fact known: $\mathbf{v}$ is assumed only to contain random errors, which must by definition have an average value of zero. The summation in Equation 11.21 can be evaluated for all the elements in $\mathbf{v}$ using Equation 11.12—noting that $\mathbf{x}$ is now zero, we can write:

$$\sum_{i=1}^{m} v_i^2 = \mathbf{v}^T \mathbf{v} = (\mathbf{R} \mathbf{k})^T (\mathbf{R} \mathbf{k}) = \mathbf{k}^T \mathbf{R}^T \mathbf{R} \mathbf{k} = \mathbf{k}^T \mathbf{W} \mathbf{k} \qquad 11.22$$

Note that the final step in this equation can be made because $\mathbf{R}$ is a diagonal matrix (i.e., $\mathbf{R} = \mathbf{R}^T$), and $\mathbf{W} = \mathbf{R} \times \mathbf{R}$.

However, although the vector $\mathbf{v}$ does indeed contain m elements, they cannot be considered to be fully independent of each other, because the n variables contained in $\mathbf{x}$ have already been deliberately used to make

$$\sum_{i=1}^{m} v_i^2$$

as small as possible. Instead of dividing by m, therefore, we should divide by $(m - n)$ to give:*

$$\sigma_v = \sqrt{\frac{\mathbf{k}^T \mathbf{W} \mathbf{k}}{m - n}} \qquad 11.23$$

If the standard deviations in the initial readings were estimated correctly, then the RHS of Equation 11.23 (commonly known as the estimated standard deviation [ESD] scale factor) will evaluate to around unity, once the least-squares iteration has converged. If the factor is greater than unity, it indicates that the estimates appear to have been on the optimistic (i.e., small) side, given the residual values. A value less than unity suggests either that the estimates were generally on the pessimistic (i.e., large) side, or that the framework of observations is poorly conditioned (i.e., not stiff enough), allowing the standard deviation of the residuals to be brought artificially low. Note, however, that the ESD scale factor is only an overall measure of the estimates—even if the value is unity, it is possible that some ESDs were too large, and some too small. For each individual reading, though, the quoted ESD is known as the *a priori* estimate of its standard deviation, and its ESD multiplied by σ_v is known as the *a posteriori* estimate of its SD.

Assuming that the relative sizes of the ESDs are more or less correct, the best estimate for the actual standard deviation of the population in $\mathbf{v}$ is now given by σ_v, and the calculation can proceed to the next stage.

If two variables, x_1 and x_2, can both be expressed as linear functions of m further variables, we can write this in the form:

---

* A detailed justification of this is given in Wolf and Ghilani (1997). A rough-and-ready justification is to consider the case when $m = n$, i.e., the number of measurements is equal to the number of variables, and there is no redundancy. Then, the variables can be adjusted until $\mathbf{v} = 0$, even if errors are present. Under these circumstances, we would expect the expression for σ_v^2 to yield an indeterminate result.

$$
\begin{bmatrix} x_1 \\ x_2 \end{bmatrix} = \begin{bmatrix} c_{11} & c_{12} & \cdots & c_{1m} \\ c_{21} & c_{22} & \cdots & c_{2m} \end{bmatrix} \begin{bmatrix} v_1 \\ v_2 \\ \vdots \\ v_m \end{bmatrix} + \begin{bmatrix} d_1 \\ d_2 \\ \vdots \\ d_m \end{bmatrix} \quad \text{or } \mathbf{x} = \mathbf{C}\,\mathbf{v} + \mathbf{d} \quad 11.24
$$

where $\mathbf{v}$ contains the m variables, and $\mathbf{C}$ and $\mathbf{d}$ contain constants and/or zeroes. Basic statistics defines the *variance* of a variable to be the square of its standard deviation, and tells us that if the standard deviation of the ith element in $\mathbf{v}$ is $\sigma_v i$, and all the variables in $\mathbf{v}$ are independent of one another, then the variance of x_1 will be:

$$
\sigma_{x_1}^{\,2} = c_{11}^{\,2}\,\sigma_{v_1}^{\,2} + c_{12}^{\,2}\,\sigma_{v_2}^{\,2} + \ldots + c_{1m}^{\,2}\,\sigma_{v_m}^{\,2} \quad\quad 11.25
$$

with x_2 having a similar expression. In addition, we must take into account the fact that statistical variations in x_2 as a result of variations in $\mathbf{v}$ will be partly coupled to variations in x_1 (and vice versa), if both are functions of the same elements in $\mathbf{v}$. This is expressed by means of a *covariance* between the two variables, which is defined by:

$$
\sigma_{x_1 x_2} = \sigma_{x_2 x_1} = (c_{11}\,c_{21})\,\sigma_{v_1}^{\,2} + (c_{12}\,c_{22})\,\sigma_{v_2}^{\,2} + \ldots + (c_{1m}\,c_{2m})\,\sigma_{v_m}^{\,2} \quad 11.26
$$

Bearing in mind that the standard deviation of each variable in $\mathbf{v}$ has been estimated to be σ_v, we can thus set up a *variance/covariance matrix* for x_1 and x_2, as follows:

$$
\begin{bmatrix} \sigma_{x_1}^2 & \sigma_{x_1 x_2} \\ \sigma_{x_2 x_1} & \sigma_{x_2}^2 \end{bmatrix} = \begin{bmatrix} c_{11} & c_{12} & \cdots & c_{1m} \\ c_{21} & c_{22} & \cdots & c_{2m} \end{bmatrix} \begin{bmatrix} c_{11} & c_{21} \\ c_{12} & c_{22} \\ \vdots & \vdots \\ c_{1m} & c_{2m} \end{bmatrix} \sigma_v^2 \quad 11.27
$$

or, in a more compact form,

$$
\sigma_x = \mathbf{C}\,\mathbf{C}^{\mathrm{T}}\,\sigma_v^2 \quad\quad 11.28
$$

In the context of least-squares adjustment, $\mathbf{x}$ is an n-dimensional vector of variables which we can express in terms of the variables in $\mathbf{v}$ by re-arranging Equation 11.12:

$$
\mathbf{R}\,\mathbf{A}\,\mathbf{x} = \mathbf{v} - \mathbf{R}\,\mathbf{k} \quad\quad 11.29
$$

whence

$$x = (R\ A)^{-1}v - (R\ A)^{-1}(R\ k) \qquad 11.30$$

The second term on the RHS is constant, so by comparing Equation 11.24 with Equation 11.30, we can see that C equates to $(R\ A)^{-1}$ in the context of least-squares adjustment. We can thus rewrite Equation 11.28 in the form:

$$\sigma_x = (R\ A)^{-1}\ [(R\ A)^{-1}]^T\ \sigma_v{}^2 \qquad 11.31$$

Elementary matrix identities then allow us to write:

$$\sigma_x = A^{-1}\ R^{-1}\ [A^{-1}\ R^{-1}]^T\ \sigma_v{}^2 = A^{-1}\ R^{-1}\ (R^{-1})^{\ T}\ (A^{-1})^T\ \sigma_v{}^2$$

$$= A^{-1}\ R^{-1}\ (R^T)^{-1}\ (A^T)^{-1}\ \sigma_v{}^2$$

$$\therefore \sigma_x = (A^T\ R^T\ R\ A)^{-1}\ \sigma_v{}^2 = (A^T\ W\ A)^{-1}\ \sigma_v{}^2 \qquad 11.32$$

Since the term $(A^T\ W\ A)^{-1}$ has already been evaluated for the final adjustments in x (Equation 11.20), the variance/covariance matrix for x is easy to compute.

The variance/covariance matrix given by Equation 11.32 contains the covariances between all the elements in x. Generally, we are only interested in the covariance between the Easting and Northing at one of the points which we are trying to fix, such as that shown in the example (Figure 11.3). A typical 2×2 matrix fragment for such a point (point i, say) could be written as:

$$\sigma_{iEN} = \begin{bmatrix} \sigma_{iE}^2 & \sigma_{iEiN} \\ \sigma_{iNiE} & \sigma_{iN}^2 \end{bmatrix} \qquad 11.33$$

One standard deviation in the east or north direction is given by σ_{iE} and σ_{iN} respectively. To find the size of one standard deviation in any other direction, we need some further statistics. If two quantities x_1 and x_2 are functions of two variables v_1 and v_2 which are not wholly independent of each other, then the variance of x_1 is given by:

$$\sigma_{x_1}{}^2 = \left(\frac{\partial x_1}{\partial v_1}\right)^2 \sigma_{v_1}{}^2 + \left(\frac{\partial x_1}{\partial v_2}\right)^2 \sigma_{v_2}{}^2 + 2\left(\frac{\partial x_1}{\partial v_1}\right)\left(\frac{\partial x_1}{\partial v_2}\right)\sigma_{v_1 v_2} \qquad 11.34$$

where $\sigma_{v_1 v_2}$ is the covariance between v_1 and v_2; and the covariance between x_1 and x_2 is given by:

$$\sigma_{x_1x_2} = \left(\frac{\partial x_1}{\partial v_1}\right)\left(\frac{\partial x_2}{\partial v_1}\right)\sigma_{v_1}^2 + \left(\frac{\partial x_1}{\partial v_2}\right)\left(\frac{\partial x_2}{\partial v_2}\right)\sigma_{v_2}^2 \qquad 11.35$$

$$+ \left\{\left(\frac{\partial x_1}{\partial v_1}\right)\left(\frac{\partial x_2}{\partial v_2}\right) + \left(\frac{\partial x_1}{\partial v_2}\right)\left(\frac{\partial x_2}{\partial v_1}\right)\right\}\sigma_{v_1v_2}$$

On the EN plane in Figure 11.10, we can write the expression for a movement in the U direction, which makes an angle θ with the E-axis, as:

$$\delta_U = \delta_E \cos\theta + \delta_N \sin\theta \qquad 11.36$$

whence

$$\left(\frac{\partial U}{\partial E}\right) = \cos\theta \text{ and } \left(\frac{\partial U}{\partial N}\right) = \sin\theta \qquad 11.37$$

Rewriting Equation 11.34 with U, E and N in place of x_1, v_1 and v_2 gives us the variance in the U direction:

$$\sigma_U^2 = \cos^2\theta\,\sigma_E^2 + \sin^2\theta\,\sigma_N^2 + 2\sin\theta\cos\theta\,\sigma_{EN} \qquad 11.38$$

Equations 11.34 and 11.35 can likewise be used to express the variance in the V direction, and the covariance between the U and V directions. These results can be combined and expressed in the following compact form:

$$\begin{bmatrix} \sigma_U^2 & \sigma_{UV} \\ \sigma_{UV} & \sigma_V^2 \end{bmatrix} = \begin{bmatrix} \cos\theta & \sin\theta \\ -\sin\theta & \cos\theta \end{bmatrix} \begin{bmatrix} \sigma_E^2 & \sigma_{EN} \\ \sigma_{EN} & \sigma_N^2 \end{bmatrix} \begin{bmatrix} \cos\theta & -\sin\theta \\ \sin\theta & \cos\theta \end{bmatrix} \quad 11.39$$

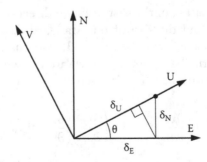

Figure 11.10 Change of axes.

Readers who are familiar with eigenvectors will now realise that there will be one orientation of the UV axes (i.e., one value of θ) for which σ_{UV} will be zero, with σ_U^2 larger than σ_V^2. At that orientation, U and V represent the principal directions of the variance/covariance matrix for the point, with the maximum possible variance occurring in the U direction and the minimum in the V direction. The sizes of these principal variances are of course the two eigenvalues of the (symmetric) σ_{EN} matrix, with the U and V directions being given by the corresponding (orthogonal) eigenvectors. Alternatively, the sizes and directions of the principal variances can be obtained from a Mohr's circle construction, similar to that used in 2-D stress calculations[*]—with variances in place of plane stress, and covariances in place of shear stress.

One standard deviation in any particular direction is of course given by the root of the variance in that direction; and normally, an engineer would simply be concerned to ensure that the largest standard deviation (i.e., the root of the largest eigenvalue of σ_{EN}) was adequately small for the job in hand. If needed, though, the standard deviations in other directions can be found: either by using Equation 11.38, or via a Mohr's circle, or more visually by means of the graphical construction shown in Figure 11.11. Here, an ellipse has been drawn on the UV axes, with semi-major and semi-minor axes of σ_U and σ_V, respectively. This is the so-called *error ellipse* for the point. The size of one standard deviation in any direction is found by proceeding along a line in that direction until a perpendicular can be

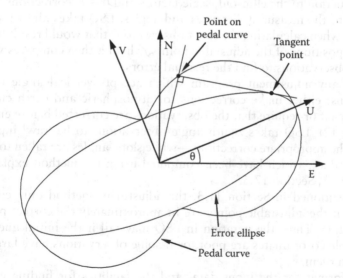

Figure 11.11 Error ellipse and pedal curve.

[*] Also used for curvature, as described in Chapter 8, Section 8.6.1.

dropped which is tangential to the error ellipse. The resulting locus of points is known as the *pedal curve* of the error ellipse. As can be seen, the largest standard deviation is given by the major semi-axis of the error ellipse—but note the standard deviation in almost every other direction is very nearly as large as this.

11.5 LEAST-SQUARES ADJUSTMENT BY COMPUTER

The mathematical methods described in the preceding two sections are clearly more suited to computer programs than to hand calculation—though it is worth remembering that the adjustment of the first-order control points in OSGB36 was done by hand, using 10-figure logarithm tables and mechanical adding machines to multiply the numbers!

Several programs are available for the least-squares adjustment of both conventional and GNSS surveying data; one such is LSQ, which was developed in the Cambridge University Engineering Department for student use, and is included with this book.

LSQ is essentially a 'batch' program which processes an input file of control points and observations, to produce a results file of residual errors and likely positions for the 'adjustable' points. The observations are all processed in a grid system, as opposed to being converted to geodetic co-ordinates for processing, and back to grid co-ordinates afterwards. This means that reduction to the ellipsoid, scale factors and $(t - T)$ corrections must be applied to the measured distances and angles; LSQ takes all of these into account when calculating the exact observations that would result from the current positions of the adjustable points, which it then compares with the actual observations to find the residual errors.

Also, any adjustment program which accepts vertical angle observations must either make corrections for atmospheric and earth curvature effects itself or require that the observations are corrected before entry (see Chapter 12). LSQ takes zenith angle observations to be 'raw' input, and makes the appropriate corrections—while slope angles are taken to be corrected values, which have been computed using the method explained in Chapter 12, Section 12.5.

As mentioned in Section 11.3, the adjustment method only converges reliably if the adjustable points are in approximately the correct positions at the start. Thus, the iteration in LSQ may fail if the initial guesses for adjustable co-ordinates are poor, or if some observations have large gross errors in them.

The format for the input data, and the facilities for finding errors in the data and maximising confidence in the results, are fully explained in the user manual, which forms the help system for the program. A sample adjustment using LSQ is given in Appendix E.

11.6 INTERPRETING LEAST-SQUARES RESULTS

The results of any least-squares adjustment should be inspected carefully before they are accepted. In particular, the following points should be checked:

1. Check that the standard deviation of the weighted residual errors (the ESD scale factor in LSQ parlance) is reasonably close to unity. If it is greater than about 1.5, it means you have either been excessively optimistic in your estimate of likely standard deviations for the readings, or that there is at least one nonrandom error amongst the data. Try eliminating the observation which has the largest weighted error, and running the adjustment again. If everything works well this time (and there are still enough readings to provide redundancy), it means that there was almost certainly a gross error of some kind in that observation.

2. If the ESD scale factor is less than about 0.7, then you may have been overly pessimistic in your estimates of the likely error present in each reading. Alternatively, it may be that there are not enough suitable readings to fix the unknown points with sufficient certainty: a large number of adjustable points which are only loosely tied into two or three distant known points may produce a network which appears adequately stiff but which is not, in fact, very well constrained. There is always a statistical possibility that the random errors have all come out close to zero. However, the probability that errors are present but are hidden because they appear to 'cancel out' will generally be greater.

3. If the ESD scale factor is around unity, check that the distribution of the weighted errors is similar to what would be expected from a normal* distribution. Typically, two-thirds of the weighted errors should be less than unity, about one in 20 should be greater than 2, and virtually none greater than 3. If the distribution differs significantly from this profile, further investigation is called for.

4. If the program has the facility, check that there is still redundancy in the remaining data—but note that it is possible to falsely simulate redundancy, e.g., by including some observations twice in the dataset.

5. If the program has no redundancy check, then check that no weighted error is exactly zero. If one or more is, then it almost certainly means that there is no other measurement in the data to provide an independent check of the quantity which it is measuring—in other words, there is no redundancy in that part of the scheme.

6. Finally, check that the sizes of the error ellipses are acceptable. If an error ellipse has a large eccentricity (i.e., it is long and thin), it indicates that the point is well fixed in one direction but poorly fixed

---

* In the statistical sense of the word.

in another (see the example in Chapter 2, Section 2.3). Further (and different) observations will then be needed to improve matters and make the error ellipse more circular. Note that the *shape* of the error ellipses is principally determined by the geometry of the observation scheme, rather than by the accuracy of the measurements themselves. The shapes of the ellipses can in fact be found without taking a single measurement, provided the likely relative accuracies of angle and distance measurements are known. Thus, the best set of measurements to produce near-circular error ellipses can (and should) be determined at the planning stage; a highly eccentric ellipse in the results probably means that the planning was poorly done.

11.7 SUMMARY

Least-squares adjustment is a powerful tool for the surveyor, but needs to be used with care and skill to produce reliable results. A few final guidelines are:

1. Make sure you have planned to take enough observations to fix each and every variable in the adjustment with some degree of redundancy— simply having more observations than unknowns is not necessarily sufficient.
2. If possible, use the adjustment package *before* any measurements have been taken, to ensure that the intended measurements will produce well-shaped (i.e., near-circular) error ellipses. Most least-squares programs have the ability to be used in this 'planning' mode; the details of how to do this will depend on the package used.
3. Most least-squares programs will require initial guesses for the positions of the 'unknown' points. Make reasonably precise initial guesses for these, such that the initial calculated angles will not differ from the observed angles by more than about 20°. If this is not done, the iteration may not converge.
4. Take care to check how the program interprets a 'horizontal distance'—this might be a reduced distance, or a distance measured at some altitude above the ellipsoid. LSQ, for instance, assumes that any horizontal distance is at the altitude of the observing station, i.e., that it has been calculated as suggested by Equation 10.5. (To make accurate use of such data, LSQ requires that the orthometric height of the observing station and the geoid-ellipsoid separation are both correct to within about 5 metres; see Chapter 10, Section 10.1 for further details.)
5. Be careful when including slope distances in the input data. If the heights of the stations at each end of the ray are fixed, then the height difference must be sufficiently accurate to allow the program to 'fit' the reduced distance into the network without needing inappropriate

adjustments in the horizontal plane. If the height of one or both stations is adjustable, and not appropriately constrained in some other way, then this freedom will allow the program to accommodate small errors in the measured slope distance by making large changes in the height difference. This is particularly problematic when the slope distance is very nearly horizontal—under these conditions, the behaviour becomes analogous to buckling. One practical solution to this problem is always to combine a slope distance measurement with a slope angle measurement, as indicated in Figure 12.5.

6. Be careful not to exclude any observation for no better reason than that it appears to disagree with your other observations. Being selective in this way can lead to a totally false indication of accuracy in the final result. It is, of course, perfectly acceptable to investigate such an observation further, to see whether there was some reason why it may not have been as accurate or reliable as the other observations. If you already have another measurement of the same observation which fits the other observations better, then it is probably safe to reject the problematic one; if you do not, it is good practice to go and measure it again.

7. Sometimes, the quality of data for the 'known' points may not be as good as you have been led to expect, and this may result in some of your observations appearing to disagree with each other. If you suspect this to be the case, and you have enough readings, you may be able to identify possible errors in the known data by making the known points 'stiff' rather than 'rigid', and seeing whether the quality of the result suddenly improves. Again, further observations may be necessary to confirm this to an acceptable level of confidence.

8. Be careful not to assign small ESDs to observations which do not warrant them. If they do not fit with other observations which have larger ESDs, then the adjustment process will load the residual error onto those other observations, because this is the 'cheapest' way to resolve the discrepancy in terms of weighted residual errors. This will give the impression that the readings with large ESDs (which may be perfectly good) are all at fault, and the reading with the low ESD (which may be poor) is correct.

9. Check all results against the criteria listed in Section 11.6 before accepting them.

Chapter 12

Trigonometric Heighting

12.1 INTRODUCTION

The height differences between control points are often explicitly required in engineering surveying work. Even when they are not, they must, for instance, be found before distance measurements can be used in accurate surveying work, as shown in Chapter 10.

The methods for measuring height differences covered earlier in this book are levelling (for short distances, or for maximum accuracy up to about 25 km), and GNSS for longer distances, or for shorter distances not requiring the accuracy available from levelling.

GNSS gives the absolute heights of stations to a more than adequate accuracy for the processing of distance measurements, but the relative heights may not be sufficiently accurate if the ray between the two stations has a steep slope (see Chapter 10, Section 10.4). Other issues which may influence the use of GNSS for height differences are:

1. GNSS may not work near buildings, in excavations, or beneath tree canopies—and it certainly won't work in tunnels.
2. Height information from GNSS is inevitably less accurate than horizontal positioning information. Differential GNSS can measure height differences to about 1 cm per kilometre of separation, but fairly long observation periods are required to achieve this.
3. The conversion of GNSS height differences to differences in orthometric height requires an accurate and reliable geoid model.*
4. There is no fully independent check to ensure that results from GNSS observations are correct. Redundancy can be achieved by making additional GNSS observations, but all the results are subsequently

---

* This is not a problem in the context of processing distances, as ellipsoidal heights are required for this purpose—but a transform may be required if an ellipsoid other than WGS84 is being used.

processed in the same way. Any systematic errors (e.g., in the antenna offsets, the transformation parameters, or the geoid model) are likely to remain undetected.

Conventional surveying methods provide a solution to some of these problems, but conventional levelling over long distances or large height differences is extremely time-consuming. In addition, some height differences are simply unsuitable for measurement by levelling: the height of a tall building or a cliff face, for instance. One solution for such problems is trigonometric heighting, which finds height differences by observing along a line of sight which is not horizontal (as in levelling) but sloped.

12.2 METHODS FOR TRIGONOMETRIC HEIGHTING

In its simplest form, trigonometric heighting consists of sighting a total station at a distant target and letting the instrument use the distance and vertical angle to compute a height difference, using Equation 10.9 from Chapter 10 (or something similar). As discussed in Section 10.4, though, the resulting height difference may be in error by up to 10 cm per kilometre of separation, due to the difficulty of measuring the vertical angle reliably.

A solution to this difficulty is to measure the vertical angle from both stations simultaneously—as can be seen from Figure 12.1, the mean slope angle[*] of the light ray (σ) stays almost constant regardless of how much the light path bend as it travels through the atmosphere.[†] As will be shown later, the value of σ can be found quite easily from the two zenith angles z_A and z_B—so most of the atmospheric effects can be eliminated if these two angles are both known. The use of reciprocal vertical angles (or RVs), as such measurements are called, can therefore enable the height difference between two stations to be calculated to a high degree of accuracy, with an observation time which is short by comparison with other available methods.

12.3 PROCEDURE FOR RECIPROCAL VERTICAL MEASUREMENTS

To take RV measurements between two stations, total stations or theodolites are set up over each station; and for maximum accuracy, targets are set up on auxiliary stations, a short distance from each instrument. The two

---

[*] The slope angle of even a straight line changes along its length because of the curvature of the earth. The mean slope angle of a ray is defined as the slope angle at its midpoint.

[†] This assumes that the curvature of the light path is approximately constant, which is usually (but not always) the case.

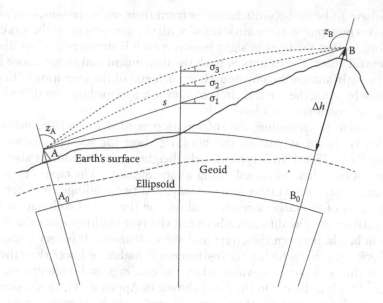

Figure 12.1 Slope angles for different light paths.

targets are set up such that the bearing from each instrument to its nearby target is perpendicular to the bearing between the two instruments. Both these offsets are also in the same direction, so that the two lines of sight between each instrument and its distant target cross each other, as shown in plan in Figure 12.2.

Each auxiliary station is set up by observing the distant main station approximately from the local main station (an accuracy of 1 minute is sufficient), then swinging the instrument through 90° and setting the auxiliary station between 2 and 4 metres away* on the given line of sight. If the layout of the land near the two main stations means that the auxiliary sta-

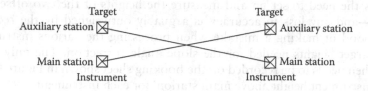

Figure 12.2 Equipment layout for reciprocal vertical angle observations.

---

* The distance should be small enough for the two lines of sight in Figure 11.1 to pass through the same body of air, and for the instrument-to-target distances to be very similar to the instrument-to-instrument distance. However, it needs to be large enough for the nearby instrument to be able to focus on the target, and to avoid the danger of the observer knocking into the target.

tions have to be on opposite bearings from their main stations, one of the targets is now put on the main station, with the instrument on the auxiliary station, so that the lines of sight between each instrument and its distant target still cross. The heights of both the instrument and target above their nearby main station are now found, at each end of the operation. This can be done by using the vertical circle to set the telescope horizontal, and then using the instrument as a level.

A scheme for recording the information needed to find the necessary heights is shown as part of the booking sheet for RV observations, in Appendix G, Figure G.7. Typically, the height of the instrument above the main station is first measured, using a tape measure. The tape is then held vertically against the target in some way, and two readings are taken: one at the level of the target's centre, and one on the line of collimation from the instrument. The difference between the two readings gives the difference in height between the target and the instrument. It is very important to check explicitly whether the instrument is higher or lower than the target, as this will not be obvious when the readings and computations are checked back at base. On the form shown in Appendix G, this is done by filling in the 'target higher' or 'target lower' section, as appropriate. Two boxes are provided for each reading, so that height measurements can be taken both before and after the main observations. It is also good practice to use face 1 for the initial measurement and face 2 for the final one—when the two readings are averaged, this will cancel our any vertical circle error, as described in Chapter 4, Section 4.4. Given the small distance involved, though, this error is unlikely to result in a difference of more than 1 or 2 millimetres—so if the difference is greater than this, there is probably another cause, e.g., the telescope was not set exactly horizontal for one of the readings, or the offset target has moved during the main observations.

If time is at a premium and visibility is good, then RV measurements can be speeded up by sighting each instrument onto the objective lens of the other instrument's telescope, rather than using an offset target. This avoids the need to set up and measure the heights of the two offset stations—and any loss in accuracy is arguably outweighed by the reduced likelihood of making a mistake when processing the various instrument and target heights needed for the slope angle correction. The only value that then needs to be recorded on the booking sheet shown in Figure G.7 is the 'Instrument height above main station' for each instrument.

Simultaneous vertical angle observations are now taken by each total station to the distant target. Greatest accuracy is achieved if all the readings are truly simultaneous, as the curvature of the light path can change quite quickly, especially during intermittent sunshine. This goal is best achieved by having an observer and booker at each station, with the bookers in radio contact. As the agreed time for the first observation approaches, each observer should ensure that his or her instrument is on face 1, and that the sighting of the telescope is approximately correct. (If appropriate, the vernier setting should also be approximately correct, and the alidade bubble should be set exactly—this will reduce the time between the two readings, which can be advantageous.) At the end of a short count-down (e.g., '3 – 2 – 1 – go') from one booker, each observer makes the final exact adjustment to the vertical tangent screw, and takes the first measurement. This is recorded by the booker, together with the number and time of the observation. Each instrument is then transited to face 2 as quickly as possible, and the telescope is again aimed at the target. When one station is ready to measure again, the booker transmits a message such as "Station B is ready for face 2" over the radio—and when the other station is also ready, the booker at that station immediately initiates a countdown for the face 2 measurement.

For best accuracy, this entire process should then be repeated four more times (typically at about 5-minute intervals) to obtain a total of five (face1 + face2) observations. For convenience, observations 2 and 4 can start on face 2, as this will be the configuration of the instrument after the previous observation—but for reasons explained below, it is important that both instruments always use the same face at the same time.

Each observation should be processed as soon as it has been made. The face 2 zenith reading is subtracted from 360°, and the corresponding face 1 reading is then subtracted from this (see the booking sheet in Appendix G). The result should be in the range ±30 seconds—this seemingly large value is acceptable as it includes any collimation and vertical circle errors on the instrument, which exactly cancel out when the average of the two sightings is computed.[*] Under ideal conditions, this difference will remain constant for each observation, but in practice it will probably vary by up to 10 seconds—either because of observation error, or because of a change in atmospheric conditions between the face 1 and face 2 readings (e.g., if one was done in full sunlight and the other when the sun was behind a cloud.)

After five observations have been completed, one team should pass their five differences over to the team at the other station, who should subtract them from their own corresponding differences. If both instruments have always read the same face at the same time, the resulting five 'differences of differences' (final column in the booking sheet at Appendix G) will be less affected[†] by any atmospheric changes during the course of an observation—so any remaining variation in the values is more likely to be due to observer error.

---

[*] A higher value suggests that the instrument requires servicing, but should not affect the accuracy of the final result.

[†] If the light always travels along a circular path, the atmospheric effects will cancel exactly. The effects of a noncircular path are reduced by minimising the time between the two readings of an observation.

If the five values do not differ by more than about 5 seconds, then an acceptable result has probably been achieved. If the spread is greater than this, then further observations should be taken until there are five observations which meet this criterion.

When enough acceptable observations have been obtained, the heights of the instrument and target at each station are remeasured, using the same procedure as before (but with the instrument on the other face). A slope distance might then be measured as well, using a total station above one main station, and a reflector above the other. This measurement will be useful for converting the instrument-to-target vertical angles into station-to-station values; but it will probably also be used to find the reduced distance or slope distance between the two stations, as described in Chapter 10, Sections 10.4 and 10.5.

12.4 SCHEME OF OBSERVATIONS

Reciprocal vertical angles can be taken between just two stations, as described above; the scheme of observations is designed to ensure that any gross error is unlikely to pass unnoticed. However, it is strongly recommended that a 'closed bay' of observations is taken, e.g., between stations A and B, B and C, then C and A. The vertical closure of this bay will give a good indication of the accuracy of the observations: if the three stations form an approximately equilateral triangle, their relative heights from RV measurements can be compared with GNSS results to verify the quality of the transform parameters used for the latter.

More elaborate schemes of RV observations can also be devised, similar to the schemes for levelling shown in Chapter 6. In addition, RV results can be freely mixed with other levelling results, to provide a fully redundant scheme of vertical control.

12.5 CALCULATIONS

The first stage of the calculation is to find the amount by which the vertical angles measured from each instrument must be altered to allow for the height of the instrument above the local station, and the height of the target above the remote station. This correction depends only on the *difference* between these heights; if the two heights were the same, no correction would need to be made. Note, however, that a different correction must be made for the two sets of observations: if offset targets have been used as described above, the height differences in the two observed rays will not be the same.

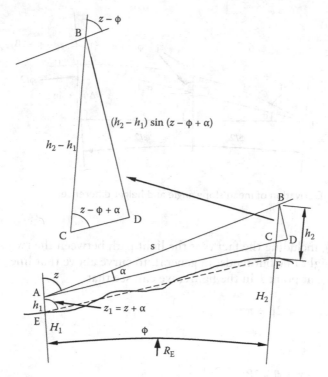

Figure 12.3 Correction of measured vertical angles.

As shown in Figure 12.3, the correction α which must be added onto each zenith angle measurement is given by the formula:

$$\sin\alpha = \frac{(h_2 - h_1)\sin(z - \phi + \alpha)}{s} \qquad 12.1$$

However, because z is in the region of $\pi/2$, and α and ϕ are both very small, this formula simplifies to:

$$\alpha = \frac{(h_2 - h_1)\sin z}{s} \qquad 12.2$$

where α is of course in radians. Note that s need not be especially accurate in this equation, since α is small: a simple instrument-to-target distance will be perfectly accurate enough, so the corrections described in Sections 10.3 and 10.5 of Chapter 10 need not be applied at this stage.

Once the observed zenith angles have been corrected by this amount, the mean slope angle of the line between the two stations can be calculated. Figure 12.4 shows the geometry of the situation between the two stations

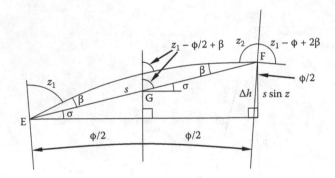

Figure 12.4 Derivation of mean slope angle and height difference.

(*E* and *F*), including the fact that the light path between the two stations is not a straight line but tends, in general, to curve above that line.

Looking at point *F* in the figure, we can see that

$$z_2 + z_1 - \phi + 2\beta = \pi \qquad\qquad 12.3$$

whence

$$z_2 + z_1 = \pi + \phi - 2\beta \qquad\qquad 12.4$$

Now looking at point *G*, we can write:

$$\sigma = \pi/2 - (z_1 - \phi/2 + \beta) \qquad\qquad 12.5$$

which, on substituting from Equation 12.4, gives:

$$\sigma = \frac{z_2 - z_1}{2} \qquad\qquad 12.6$$

A mean value of σ can therefore be found by taking an average of all the sets of readings, and it can be seen that the difference in heights between the two stations is given by:

$$\Delta h = \frac{s\sin\sigma}{\cos(\phi/2)} \approx s\sin\sigma \qquad\qquad 12.7$$

This final approximation is justified, since ϕ is generally very small; if *s* is less than 20 km, then ϕ will be less than 3×10^{-3} radians, and $\cos(\phi/2)$ will differ from unity by less than 1 part per million.

Equation 12.7 can be used to calculate the $(H_2 - H_1)$ term needed in Equation 10.38. It is not necessary to use the station-to-station slope distance for this (and that probably won't be available yet anyway, if Equation 10.38 is being used)—the instrument-to-target distance will provide sufficient accuracy for the calculation to proceed.

Note that the curvature of the earth *and* the curvature of the light path have cancelled out in Equation 12.6, as a result of measuring the zenith angle at both ends of the ray. It should be borne in mind that this only works if the curvature of the light path is constant, or nearly so; if the ray grazes the ground at any point, for instance, the accuracy of Equation 12.6 may be considerably reduced.

Assuming that the light path does have constant curvature, however, we can use the value of β computed from Equation 12.4 to estimate a value for the refraction constant defined in Equation 10.23. Comparing Figures 10.5 and 12.4, we can see that the angle labelled z_1' in Figure 10.5 is equal to $(z_1 - \phi + 2\beta)$ in Figure 12.4. So by substituting for z_1' in Equation 10.20 and then using the approximation of Equation 10.1 for d_R, we can write:

$$\frac{1}{R_1} = \frac{z_1 - (z_1 - \phi + 2\beta)}{d_R} \approx \frac{\phi - 2\beta}{s \cos \sigma} \qquad 12.8$$

Now applying Equation 10.23 to the LHS and Equation 12.4 to the RHS, we can say:

$$\frac{1}{R_E}(1 - \kappa) \approx \frac{z_2 + z_1 - \pi}{s \cos \sigma} \qquad 12.9$$

which gives

$$\kappa \approx 1 - \frac{R_E(z_2 + z_1 - \pi)}{s \cos \sigma} \qquad 12.10$$

The value for κ obtained from Equation 12.10 can be compared with the commonly-accepted value of $1/7$ for visible light. If visible or infrared radiation has been used to measure the slope distance between the two stations as part of the same exercise, then this calculated value of κ (which derives from the actual atmospheric conditions at the time of the measurement) should be used in Equation 10.26 for the precise adjustment of that slope distance.

Finally, when an accurate station-to-station slope distance has been computed using Equation 10.38, Equation 12.7 can be used again to provide a fully accurate estimate of the height difference between the two stations. Height differences computed in this way should be used when evaluating the closure of the bays mentioned in Section 12.4.

A worksheet for processing the results of reciprocal vertical angle observations is given in Appendix H; note that the formulae have been adapted to work with angles recorded in degrees, minutes and seconds, rather than in radians. Finally, the flowchart in Figure 12.5 shows how slope distance measurements and RV angle observations can be processed together, to provide suitable data for a least-squares adjustment.

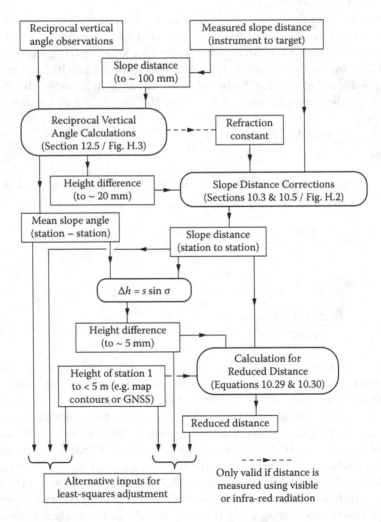

Figure 12.5 Processing reciprocal vertical angles and slope distances.

12.6 ACCURACY OF RECIPROCAL VERTICAL ANGLES

Surveying exercises at Cambridge University have now accumulated ample empirical data to show that, using instruments capable of measuring vertical angles to 1 second of accuracy, closed RV bays consisting of four control points and a total circuit length of around 5 km will reliably close to within 2 cm. This suggests an accuracy in the region of 4 mm per kilometre of separation, or an accuracy in the slope angles of about 1 second— i.e., the accuracy of the instruments used to make the observations. It is therefore possible that even higher accuracy could be achieved with better instruments, though the typical spread of results used in the lower half of the 'Calculation for Reciprocal Vertical Angles' form (Appendix H) suggests that this is unlikely.

Even with this level of accuracy, though, RV observations fill a very useful gap in precise surveying. Chapter 10 has indicated that an accuracy of 5 mm per km is required for the precise computation of reduced distances when the slope angle is around 20°, and this is certainly within the capabilities of RV observations. Differential GNSS observations are just about able to achieve this accuracy on differential height measurements (see Chapter 7, Section 7.3), but this would typically involve more expensive equipment and a longer observation time than is needed for the equivalent RV observations. Conventional levelling (especially precise levelling) can certainly provide similar or greater accuracy, but is likely to be far more time consuming—especially when, for instance, the height difference is required between two points which are on either side of a deep valley. Thus, at least for slope angles of 15° or more, RV observations are probably the most economic method of measuring height differences to the required accuracy.

Appendix A:
Constants, Formulae, Ellipsoid and Projection Data

USEFUL CONSTANTS AND FORMULAE

Approximate mean earth radius: 6.371×10^6 m
Speed of light *in vacuo*: $299.792\ 458 \times 10^6$ ms^{-1}
1 radian = $57.295\ 779\ 513°$

Sin rule: $\dfrac{a}{\sin A} = \dfrac{b}{\sin B} = \dfrac{c}{\sin C}$

Cosine rule: $c^2 = a^2 + b^2 - 2ab \cos C$

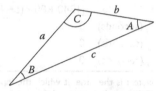

Common ellipsoids

Name	Semi-major axis, a (metres)	Reciprocal of flattening	Uses
Airy 1830	6 377 563.396	299.3249646	British National Grid (OSGB36)
Bessel 1841	6 377 397.2	299.15	Central Europe, Chile, Indonesia
Clarke 1866	6 378 206.4	294.98	North America, Philippines
Clarke 1880	6 378 249.2	293.47	Africa, France
Everest 1830	6 377 276.3	300.80	India, Burma, Afghanistan, Thailand
GRS 80 (1980)	6 378 137.0	298.2572221	North America OSTN02/OSGM02
International 1924 (Hayford 1909)	6 378 388.0	297.0	UTM
International Astronomical Union 1968	6 378 160	298.25	Australia
Krasovsky 1940	6 378 245	298.3	Russia
WGS72 (1972)	6 378 135	298.26	Oil industry
WGS84 (1984)	6 378 137.0	298.2572236	WGS84, ETRS89, ITRS

Data for transverse Mercator projections

Name	Location of true origin	False co-ordinates of true origin	Central scale factor
British National Grid	49° N, 2° W	400 000 E −100 000 N	0.999 601 272
UTM Zone 1	0° N, 177° W ⎫	500 000 E 0 N	
UTM Zone 30	0° N, 3° W ⎬	(northern hemisphere)	0.999 600 000
UTM Zone 60	0° N, 177° E ⎭	500 000 E 10 000 000 N (southern hemisphere)	

Useful conformal transforms

From	ITRF2008	ITRF2000	ETRS89
To	ITRF2000	ETRS89	OSGB36*
t_x (metres)	$-0.0019 + 0.0001 \times (t - 2000.0)$	0.054	−446.448
t_y (metres)	$-0.0017 + 0.0001 \times (t - 2000.0)$	0.051	125.157
t_z (metres)	$-0.0105 + 0.0018 \times (t - 2000.0)$	−0.048	−542.060
s	$(1340 + 80 \times (t - 2000.0)) \times 10^{-12}$	0	20.4894×10^{-6}
r_x (seconds)	0	$0.000081 \times (t - 1989.0)$	−0.1502
r_y (seconds)	0	$0.000490 \times (t - 1989.0)$	−0.2470
r_z (seconds)	0	$-0.000792 \times (t - 1989.0)$	−0.8421

Note: t is the time at which the input data for the transform was valid, expressed in decimal years. Other symbols are as defined in Equation 8.26.

* Note that this is only an approximate transform between ETRS89 and OSGB36—the definitive (but nonconformal) transform is provided by OSTN02.

Appendix B: Control Stations

WHAT IS A CONTROL STATION?

The essence of a control station is a small mark set immovably into the ground, such that an instrument (e.g., a total station or satellite receiver) or optical target can be set up above it, to an accuracy of about 1 mm in the horizontal plane.

WHERE ARE THEY PLACED?

Control stations are not usually placed in an exactly predetermined position. The normal process is to choose a location where a control station would be useful, and to place the station somewhere in that locality. Having built the station, precise measurements are then taken to determine exactly where it has, in fact, been placed.

The factors which influence the positioning of a control station are as follows

1. If it is to be used for setting out, mapping, or for deformation monitoring, then it should be placed where all relevant places and features can be easily seen, without the line of sight passing close to another object such as a building or hillside (a 'grazing ray'). If the station is to be used in conjunction with other similar stations for these purposes (as is usually the case), then the different lines of sight from the stations should form a well-conditioned shape, so that the positions of the observed points will be found to the greatest possible accuracy.

2. If the exact position of a new control station is to be fixed by conventional means, then it must be visible from at least two other control stations (and preferably from more). Sometimes, additional control stations are introduced into a network simply because they will be visible to several 'useful' stations, and will therefore improve the accuracy to which the positions of those stations are known.

3. If the station is to be used for GNSS, then a large area of sky should be visible at the station (particularly towards the equator), and there should not be any high walls nearby which might reflect satellite signals towards the receiver.

4. If an instrument is to be left unattended at a station (e.g., a motorised total station or a reference GNSS receiver), then the station must be in a secure place, such that the instrument cannot be stolen or disturbed while the surveyor is elsewhere.

5. As far as possible, a station should be sited in a place where it will be easy and safe to use (away from noise, vibration, traffic, etc.) and unlikely to be disturbed or destroyed during its anticipated useful life. Stations sited near roads or on tarmac pathways are always at risk of being covered over and lost without trace. Stations in the middle of building sites are at risk of being dug up or run over by heavy construction traffic. The latter may not destroy a station, but it could move it slightly—and thus cause all subsequent observations involving the station to be subtly inconsistent with those made beforehand.

WHAT DO CONTROL STATIONS LOOK LIKE?

The physical appearance of a control station depends mainly on the place where it is sited, and its anticipated useful lifespan. In open ground, a short-term control station might be a 1 mm diameter hole or 'centre-pop' in a brass tack driven into a short (30 cm) wooden stake, which is then hammered into the ground; on tarmac, it might be a centre mark on a stainless steel 'road bolt', which is likewise hammered into the ground. Such road bolts normally have a hemispherical head with a diameter of about 5 mm, on top of a fixed disk about 20 mm in diameter—this makes them suitable places on which to stand a levelling staff. They may also have a coloured plastic washer and/or a circle or triangle painted round them, for identification purposes.

For a more permanent marker in open ground, a precast reinforced concrete block with a suitable marker on its surface might be set into the ground, so that only its top surface is visible. Alternatively, a hole can be dug with some ferrous reinforcing bars arranged inside it, and a quantity of concrete poured in, with a nonrusting marker fixed so as to emerge slightly above the surface of the concrete when it has set: a small solid brass doorknob, some threaded steel rod which it will screw onto, a hand drill (to make a centre-pop), some ready-mix concrete for fence posts and a bucket of water is all that is required. This gives an extremely durable station at very modest cost, which has the added advantage that it can be covered over with a piece of turf or layer of soil, and thus escape the risk of being vandalised when not in use. If the upper surface of the marker is spherical, then its highest point can also conveniently be taken to be the height of the station.

A control station on a construction site would normally be surrounded by a low rectangular 'fence', made of brightly-painted wood, to warn drivers of its existence. This reduces the likelihood of the station being run over by a heavy vehicle; and a broken fence gives a helpful indication that this may have happened.

In Britain there are still many 'trig pillars' to be found on hilltops, which formed part of the conventional control network used until the advent of GNSS. Despite their robust appearance, these are intricately designed monuments and act as a housing for the actual station marker, which is near ground level inside the pillar (the height marker is a separate benchmark on the side of the pillar). In addition, there is a secondary station marker in a buried chamber directly beneath the main marker, so that the station can be recovered if the main pillar is destroyed. Such pillars are highly durable, but nonetheless needed regular inspection to detect and repair the damage, both accidental and deliberate, which they sometimes suffer.

Some of these trig pillars are nonetheless still maintained, and form part of the newer network of GNSS stations around the UK. However most passive GNSS control stations are markers set in concrete slightly below ground level, as described above.

HOW CAN THEY BE FOUND?

As implied above, the less obtrusive a control station is, the less likely it is to suffer damage. Recently-constructed stations can therefore be virtually impossible to find, unless you know exactly where to look.

When a new station is constructed, an essential part of the process is therefore to draw a small sketch-map of the area, called a *finding map*, showing clearly where the station is in relation to other recognisable features nearby. At least three measurements should be taken from the station to definite 'measurable' reference points—such as a tree, the corner of a manhole cover, a nail driven into the top of a particular fencepost, or the perpendicular distance to the edge of a nearby road. These should be taken with a tape measure, correct at least to the nearest 5 centimetres; and should be arranged such that the station could still be found even if it is well buried, and one or more of the reference points have subsequently disappeared. It is helpful to show the approximate direction of magnetic north on the sketch, too.

To find a buried station, a surveyor should ideally be equipped with a paper copy of the finding map described above, the co-ordinates of the station, a hand-held GNSS receiver, two 30-m tape measures, a metal detector, a ranging rod and a spade.

The hand-held receiver and the station co-ordinates will narrow the search to about 10 metres, such that the features on the finding map are clearly recognisable. Measuring simultaneously from two of the reference points should then indicate where to dig, and the metal detector can be used in cases of doubt, assuming the concrete contains some ferrous metal (as recommended above). If it does not, the ranging rod can be used to stab into the ground until the surface of the concrete is detected.

Appendix C: Worked Example in Transforming between Ellipsoids

This example shows how the geodetic co-ordinates of a station can be converted from one system to another, following the method given in Chapter 8, Section 8.5. In this case, the initial co-ordinates are quoted in the ETRS89 system, so are based on the WGS84 ellipsoid; and they are to be converted to the Airy ellipsoid, whose position and orientation was defined so as to conform closely with the British geoid. A transform between these two systems is published by the Ordnance Survey, and is given in Appendix A.

Note that 1 second of arc at the centre of the earth subtends about 31 metres on the earth's surface—so to preserve accuracy to 1 cm, latitude, longitude and rotation must be quoted to four decimal places of seconds.

Station:	Active GPS receiver at Ordnance Survey Headquarters, Southampton, UK
ETRS89 geodetic co-ordinates:	$\phi = 50° 55' 52.60562''$ N
	$\lambda = 1° 27' 1.85155''$ W
	$h = 100.399$ m
Data for WGS84 ellipsoid:	$a = 6378137.000$
(from Appendix A)	$r = 298.2572236$
From Equation 8.6:	$e^2 = 6.694379989 \times 10^{-3}$
From the data above:	$\sin^2 \phi = 0.6027823555$
Using Equation 8.12:	$r_N = 6391044.780$
From Equation 8.21:	$x = 4026741.601$
From Equation 8.22:	$y = -101963.784$
	λ is west of Greenwich, so $\sin(\lambda)$ is negative
From Equation 8.23:	$z = 4928807.847$
Transform parameters:	$t_x = -446.448$
(from Appendix A)	$t_y = 125.157$
	$t_z = -542.060$
	$s = 20.4894 \times 10^{-6}$
	$r_x = -0.1502'' = -728.2 \times 10^{-9}$ radians

$r_y = -0.2470'' = -1197.5 \times 10^{-9}$ radians

$r_z = -0.8421'' = -4082.6 \times 10^{-9}$ radians

From Equation 8.26: $x' = 4026371.340$

$y' = -101853.567$

$z' = 4928371.671$

From Equation 8.27: $\lambda' = 1° \ 26' \ 56.68889'' \ W$

(since y' is negative and x' is positive)

Data for Airy ellipsoid: $a' = 6377563.396$

(from Appendix A) $r' = 299.3249646$

From Equation 8.6: $e'^2 = 6.670540000 \times 10^{-3}$

Using Equation 8.31: $(1 - e'^2) = 0.9933294600$

$\sqrt{x'^2 + y'^2} = 4027659.410$

so $\phi'_1 = 50° \ 55' \ 50.60667'' \ N$

From Equation 8.32: $r'_{N1} = 6390423.710$

From Equation 8.34: $\phi'_2 = 50° \ 55' \ 50.60104'' \ N$

From Equation 8.32: $r'_{N2} = 6390423.709$

From Equation 8.34: $\phi'_3 = 50° \ 55' \ 50.60102'' \ N$

From Equation 8.32: $r'_{N3} = 6390423.709$

From Equation 8.34: $\phi'_4 = 50° \ 55' \ 50.60102'' \ N$

This has now converged, so the current values of ϕ' and r'_N can be accepted.

From Equation 8.35: $h' = 53.245$ m

Note that h' is an ellipsoidal height; the orthometric height of the station is about 0.5 m less than this, as shown for the Southampton area in Figure 8.2. As a check that the transformation has been applied in the correct direction, the original ellipsoidal height of 100.4 metres, minus the local geoid-ellipsoid separation of 47 metres shown in Figure 8.3, gives a reasonably similar orthometric height.

Appendix D: Calculation of Local Scale Factors in Transverse Mercator Projections

D.I QUICK CALCULATION

The 'quick' formula for calculating a scale factor is

$$S = S_0 \left(1 + \frac{(E - E_0)^2}{2 \times R_E^2} \right)$$

D.1

where S_0 is the central scale factor, E_0 is the false Easting of the true origin, and R_E is the mean radius of the earth.

This formula is accurate to 2 parts per million at all places within 200 kilometres of the central meridian, and to 12 parts per million up to 500 kilometres, at the latitude of the UK.

D.2 PRECISE CALCULATION

This calculation is a simplified (but no less accurate) adaptation of the formulae given in Ordnance Survey (1950). As far as possible, the same symbols have been used here, to enable comparison between the two approaches.

The data needed to start the calculation are as follows:

- the semi-major axis (a) and some other property (semi-minor axis, reciprocal of flattening, or eccentricity) of the ellipsoid;
- the central scale factor (F_0), and the false Easting of the true origin (E_0); and
- the exact Easting (E) and approximate Northing (N) of P, the point where the local scale factor is to be calculated.

1. The first step is to calculate e^2, the square of the eccentricity of the ellipsoid. If this is not directly available, it can be calculated from

$$e^2 = \frac{a^2 - b^2}{a^2} \text{ or } e^2 = \frac{2r - 1}{r^2} \qquad\qquad \text{D.2}$$

where b is the semi-minor axis of the ellipsoid, and r is the reciprocal of flattening.

2. Use E and N to estimate ϕ, the geodetic latitude of P. This needs to be estimated to the nearest $0.2°$ to achieve an accuracy of eight significant figures—but even an error of $5°$ will only affect the final answer by less than 1 part per million.

   (Strictly the value which should be used in the calculations below is ϕ', the latitude of the point on the central meridian with the same Northing as P. However, the difference between ϕ and ϕ' is never more than $0.1°$, provided the difference in the longitudes of the two points is less than $4°$—so the distinction has little practical significance.)

3. Set

$$v^2 = \frac{a^2}{1 + e^2 \sin^2 \phi} \text{ and } \eta^2 = \frac{e^2 (1 - \sin^2 \phi)}{1 - e^2} \qquad\qquad \text{D.3}$$

4. Set

$$X = \left(\frac{E - E_0}{F_0} \right)^2 \times \frac{1 + \eta^2}{v^2} \qquad\qquad \text{D.4}$$

5. The local scale factor is then given by:

$$F = F_0 \left(1 + \frac{X}{2} + \frac{X^2 (1 + 4\eta^2)}{24} \right) \qquad\qquad \text{D.5}$$

D.3 EXAMPLE

Local scale factor at Framingham, UK, on the British national grid:

Framingham is a first-order control point in the British national grid, towards the eastern edge of the projection. Its co-ordinates are quoted as 626 238 249 E, 302 646 415 N. This example has been used because it is also used as an example in Ordnance Survey (1950).

Precise calculation:
From Appendix A:

$a = 6377563.396$ $r = 299.3249646$ for the Airy ellipsoid
$E_0 = 400,000.0$ $F_0 = 0.999\ 601\ 272$ for the British national grid

1. From D.2: $e^2 = 0.006\ 670\ 540$
2. $\phi \approx 52.5°$ (to the nearest 0.1°, by estimation from an atlas). Note that the longitude, λ, of the point is about 1.3° E, i.e., about 3.3° east of the central meridian.
3. From D.3: $v^2 = 4.08448 \times 10^{13}$ $\eta^2 = 2.48864 \times 10^{-3}$
4. From D.4: $X = (5.12246 \times 10^{10}) \times (2.45439 \times 10^{-14}) = 1.25725 \times 10^{-3}$
5. From D.5: $F = 0.999\ 601\ 27(1 + 0.000\ 628\ 62 + 0.000\ 000\ 07)$
                   $= 1.000\ 229\ 7$

Note that although just six significant figures have been shown in Steps 1 to 4, this is quite sufficient to form a final answer which is correct to eight significant figures in Step 5. Note also that the final term in expression D.4 need only be included if eight or more significant figures are required.

Quick calculation:
Applying the 'quick' formula to the same example gives:

$$S = 0.99960127 \times \left(1 + \frac{(626238.249 - 400000)^2}{2 \times (6.371 \times 10^6)^2} \right) = 1.000\ 231\ 5$$

This differs from the 'accurate' value by just under 2 parts per million.

Appendix E: Worked Examples in Adjustment

E.1 BOWDITCH ADJUSTMENT

This example shows how the Bowditch calculation sheet, introduced in Chapter 11, is used in a simple traverse to fix the positions of two unknown points (*C* and *D*, in Figure E.1).

The scheme of observations is as shown in Figure E.1, with stations *A*, *B*, *E* and *F* having known co-ordinates. Note that Figure E.1 is not a scale drawing but is sketched sufficiently accurately so that the bearings are correct to within 30° or so. It is helpful to make such a sketch before starting the calculations (and even before making the observations) to guard against gross errors.

The first stage in the calculation is to transfer the initial data (i.e., the Eastings and Northings of the known stations, in the British national grid) and the observations (i.e., the measured angles and distances) into the shaded boxes on the calculation sheet, as shown in Figure E.2.

The next step is to enter the differences in Eastings and Northings for point A relative to B (*Ad−Bd* on the calculation sheet) and for point *F* relative to *E* (*Fd−Ed* on the sheet). These are used to work out the bearings

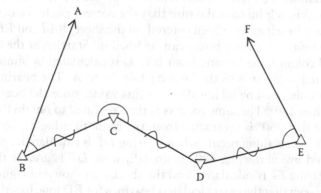

Figure E.1 Scheme of observations for a four-point traverse.

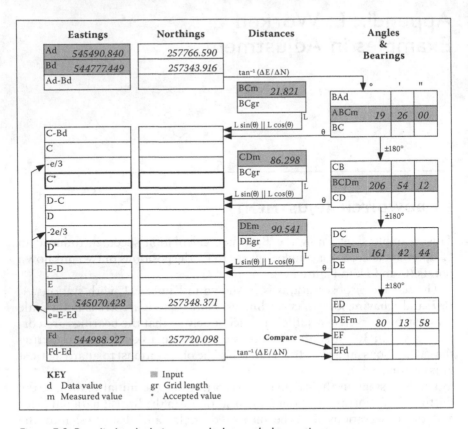

Figure E.2 Bowditch calculation, step 1: data and observations.

from *B* to *A* and from *E* to *F*. In the case of *BA*, the calculation is straight-forward, since *A* is to the north and east of *B*, as shown in Figure E.3. For *EF* the situation is more complex, and a simple sketch like the one shown in Figure E.3 is helpful to make sure that the correct angle is calculated.

Once these bearings have been entered on the sheet (*BAd* and *EFd* respectively), the remaining angle boxes can be filled in. Starting at the top of the right-hand column, the bearing from *B* to *C* is calculated by simply adding the measured angle at *B* to the bearing from *B* to *A*. The bearing from *C* to *B* is then obtained by adding 180° to this value, since the bearing *BC* is itself less than 180°. The same process is then repeated to obtain the bearing *CD*, except that 360° is subtracted from the result to bring it into the range 0–360°. The same thing occurs when bearing *DE* is calculated; and bearing *ED* is found by subtracting 180° from *DE*, since *DE* is greater than 180°. Finally, bearing *EF* is calculated, and the sheet is as shown in Figure E.4.

It can be seen that the two calculated bearings for *ED* (one based on observations and one based purely on data) differ by 7 seconds. This is reasonable

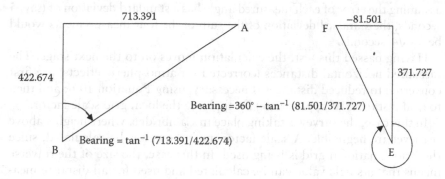

Figure E.3 Calculation of bearings from grid co-ordinates.

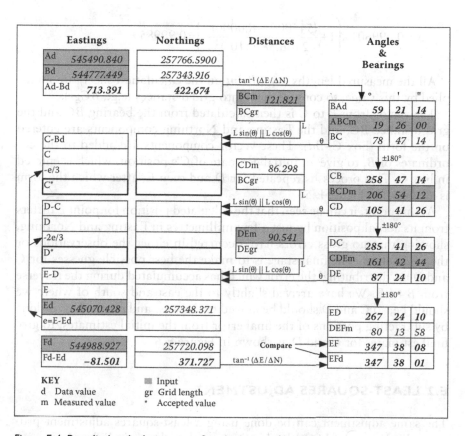

Figure E.4 Bowditch calculation, step 2: calculation of bearings.

assuming the error of each measured angle has a standard deviation of (say) 5 seconds, the standard deviation of the sum of the four measurements would be $5 \times \sqrt{4}$ seconds.

Having passed this test, the calculation moves on to the next stage. The measured horizontal distances (corrected for atmospheric effects) are first converted to reduced distances if necessary, using Equation 10.6; and then to grid distances (if appropriate) by applying the local grid scale factor.

In this case, the survey is taking place in Cambridge, where heights above sea level are negligible. A scale factor must, however, be calculated, since the British national grid is being used; in this case, the size of the traverse means that a single value can be calculated and used for all distance measurements. The survey is less than 200 km from the central meridian, and an accuracy of 2 parts per million will be more than adequate, so Equation D.1 can be used.

A suitable mean Easting is 545000 metres, and E_0 and S_0 are given in Appendix A. Putting these into Equation D.1 gives:

$$S = 0.999601 \times \left(1 + \frac{(545000 - 400000)^2}{2 \times (6.381 \times 10^6)^2}\right) = 0.999859$$

All the measured lengths (e.g., BCm on the calculation sheet) are multiplied by this value, to convert them into grid distances (e.g., $BCgr$).

The vector from B to C is then calculated from the bearing BC and the grid distance BC, and the Easting and Northing components are entered on the form (box $C–Bd$). These vector components are added to the co-ordinates of B, to give an initial estimate of C's position, which is entered in box C. This process is repeated for D and then E, after which the form is as shown in Figure E.5.

At this point, it can be seen that the calculated position for point E differs from its actual position by just a few millimetres in Easting and Northing, showing that no gross errors have occurred in either the observations or the calculation. The final stage is to make the best possible guesses for C and D by 'distributing' the error which has accumulated during the traverse from B to E. We have arrived slightly to the east and south of where we should be, so C and D should be moved to the west and north. This is done by subtracting portions of the final error from the initial estimates, to give final estimates for C and D as shown in Figure E.6.

E.2 LEAST-SQUARES ADJUSTMENT

The same adjustment can be done using a least-squares adjustment program. In the case of LSQ, the data is entered as shown in Figure E.7.

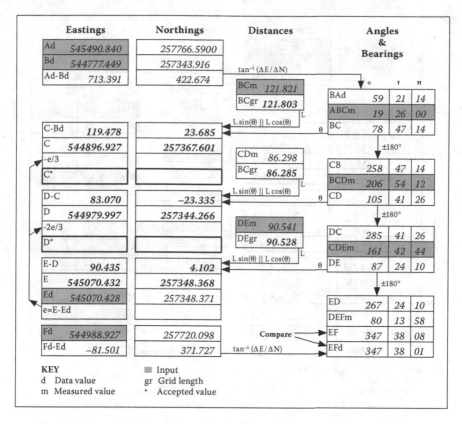

Eastings	Northings	Distances	Angles & Bearings		
				'	''
Ad 545490.840	257766.5900				
Bd 544777.449	257343.916	tan⁻¹ (ΔE/ΔN)			
Ad-Bd 713.391	422.674				
		BCm 121.821	BAd 59	21	14
		BCgr 121.803	ABCm 19	26	00
		L sin(θ) ‖ L cos(θ)	BC 78	47	14
C-Bd 119.478	23.685	θ			
C 544896.927	257367.601		±180°		
-e/3		CDm 86.298	CB 258	47	14
C*		BCgr 86.285	BCDm 206	54	12
		L sin(θ) ‖ L cos(θ)	CD 105	41	26
D-C 83.070	−23.335	θ			
D 544979.997	257344.266		±180°		
-2e/3		DEm 90.541	DC 285	41	26
D*		DEgr 90.528	CDEm 161	42	44
		L sin(θ) ‖ L cos(θ)	DE 87	24	10
E-D 90.435	4.102	θ			
E 545070.432	257348.368		±180°		
Ed 545070.428	257348.371		ED 267	24	10
e=E-Ed			DEFm 80	13	58
Fd 544988.927	257720.098	Compare	EF 347	38	08
Fd-Ed −81.501	371.727	tan⁻¹ (ΔE/ΔN)	EFd 347	38	01

KEY
d Data value   ■ Input
m Measured value gr Grid length
                    * Accepted value

Figure E.5 Bowditch calculation, step 3: preliminary grid positions.

The input file follows the rules given in the LSQ help system. The first line is a title, and is followed by a line which specifies the projection (in this case, the British national grid), to enable the local scale factor(s) to be calculated.

The next group of lines describes the control stations. Stations *A*, *B*, *E* and *F* are entered with their known Eastings and Northings, which are each 'fixed' by the letter F which follows them. Stations *C* and *D* are given approximate co-ordinates, which are set to be adjustable by the letter A. The heights of all the stations are fixed, as this is a 2-D adjustment – a reasonable height has been chosen for each station, so that LSQ will generate the correct reduced distances from the quoted horizontal distances.

The horizontal angle observations are entered next, in degrees, minutes and seconds; the final number is an estimate (in seconds) of the standard deviation of the error which might be expected in each observation. Two seconds is a typical value for an observation made under favourable conditions using a good quality instrument.

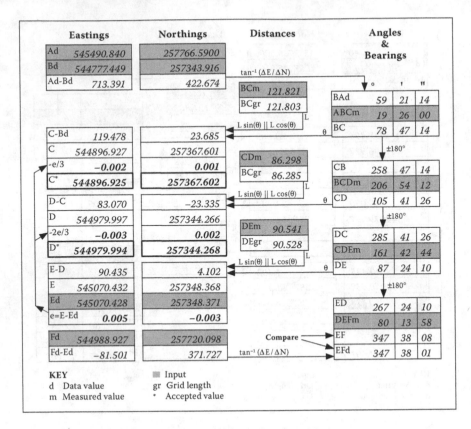

Figure E.6 Bowditch calculation, step 4: adjusted grid positions.

```
Four Point Traverse
Projection BNG

A 545490.840  F   257766.590  F   10 F
B 544777.449  F   257343.916  F   10 F
E 545070.428  F   257348.371  F   10 F
F 544988.927  F   257720.098  F   10 F
C 544800      A   257500      A   10 F
D 544900      A   257300      A   10 F

HA A B C   19 26 0  2
HA B C D  206 54 12 2
HA C D E  161 42 44 2
HA D E F   80 13 58 2

HD B C 121.821  .005
HD C D  86.298  .005
HD D E  90.541  .005
```

Figure E.7 Four-point traverse: input data for LSQ.

The last group of lines records the measured horizontal distances, in metres; again, the final number is the estimated standard deviation (ESD) of the reading. Typically, most EDM devices have a standard deviation of ±5 mm, even on short rays such as these.

Running LSQ produces the result shown in Figure E.8, after nine cycles of adjustment. It turns out that this number of cycles is necessary for the adjustment to converge, because of the relatively poor initial guess for the position of C.

The co-ordinates which LSQ has chosen for C and D are about 1 mm different from those found in the Bowditch adjustment above. These small differences arise from the different relative weightings that LSQ is able to give to angle and distance measurements, and from the fact that the Bowditch method only uses the measured angle DEF as a check, and not as part of the adjustment.

```
Four Point Traverse

* Program: LSQ V-8.80   Date: 15/05/2013    Time: 11:29:35   Cycles: 9
* Projection: BNG (British National Grid)  Ellipsoid: Airy 1830
Projection BNG  Airy  0

* Station co-ordinates (Adjustable co-ordinates updated)   Error ellipses
*        EASTING           NORTHING          HEIGHT        MAJOR   MINOR  AZ HEIGHT
A     545490.8400 F     257766.5900 F     10.0000 F
B     544777.4490 F     257343.9160 F     10.0000 F
E     545070.4280 F     257348.3710 F     10.0000 F
F     544988.9270 F     257720.0980 F     10.0000 F
C     544896.9245 A     257367.6011 A     10.0000 F   * 0.0041 0.0008 84
D     544979.9951 A     257344.2669 A     10.0000 F   * 0.0045 0.0007 88

* Analysis of readings:
* REF INST OBS      OBSERVED    ESD    HT-HI     CALCULATED    DIFF    WDIFF  ID
HA A  B    C     19 26  0.00   2.00     |     19 25 58.97   1.0307    0.52    1
HA B  C    D    206 54 12.00   2.00     |    206 54 10.44   1.5581    0.78    2
HA C  D    E    161 42 44.00   2.00     |    161 42 42.02   1.9822    0.99    3
HA D  E    F     80 13 58.00   2.00     |     80 13 55.60   2.4012    1.20    4
HD    B    C       121.8210   0.005     |       121.8180   0.0030    0.60    5
HD    C    D        86.2980   0.005     |        86.2978   0.0002    0.03    6
HD    D    E        90.5410   0.005     |        90.5389   0.0021    0.43    7

ESD Scale Factor = 1.131 after 9 cycles

Biggest errors:
WDIFF      1.20       0.99       0.78       0.60       0.52       0.43
Obs ID        4          3          2          5          1          7
Status
Rejection limit: (None)

Initial co-ordinates and shifts for adjustable co-ordinates:
STN   OLD EASTING & SHIFT   OLD NORTHING & SHIFT    OLD HEIGHT & SHIFT
C     544800.0000  96.9245   257500.0000 ********    10.0000   (n/a)
D     544900.0000  79.9951   257300.0000  44.2669    10.0000   (n/a)
```

Figure E.8 Four-point traverse: LSQ results.

Usefully, LSQ is also able to show the likely accuracy to which C and D have been found, by means of the error ellipses shown on the printout. The major semi-axis of each ellipse is 4 mm, so there is a 95% confidence that C and D lie within 8 mm of their calculated positions (assuming, of course, that the positions of the other points are error-free).

The next block of results compares the readings which were observed with those which would result from the calculated positions of the adjustable points. The differences, or residual errors, are shown both in seconds or millimetres (as appropriate), and also as weighted differences using the ESD quoted for the reading.

Below this the 'ESD scale factor' is shown. This is the calculated standard deviation of the weighted residual errors, and so indicates whether the residual errors are generally bigger or smaller than might have been expected from the quoted ESDs. An ESD scale factor of 2 or more would indicate that either the quoted ESDs were optimistically small, or that there errors other than observation errors in the data. Here, the ESD scale factor looks fine.

Appendix F: Worked Example in Setting Out

The purpose of conducting a traverse of the type described in Appendix E would typically be to establish additional local control points, in order to set out specified points for construction work

Suppose now that it is required to set out a foundation point X at the co-ordinates (544850.000 E, 257200.000 N), to high accuracy. The accepted co-ordinates of the nearby stations can be taken from the LSQ results shown in Figure E.8, and summarised as

Point	Easting	Northing
A	545490.840	257766.590
B	544777.449	257343.916
C	544896.925	257367.601
D	544979.995	257344.267
E	545070.428	257348.371
F	544988.927	257720.098

From a simple sketch map (Figure F.1) it is clear that stations B, C and D would be suitable for setting out X. It is good practice to use a station as far away as possible as a reference object, so we will use station A as a reference for B and C. To avoid the possibility of a systematic error caused by a mistranscription of A's co-ordinates, we will use station F as a reference for station D.

F.I MANUAL CALCULATION

To calculate the angle which must be turned through at B, we need the bearings BA and BX. The first of these has already been calculated, as shown in Figures E.3 and E.4. The other is calculated in a similar manner, as shown in Figure F.2.

225

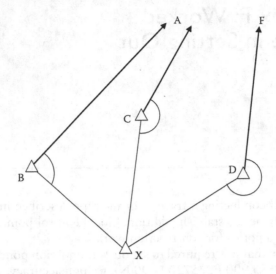

Figure F.1 Sketch map for setting out point X.

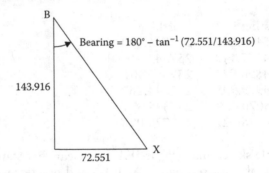

Figure F.2 Calculation of bearing from B to X.

The results can be summarised as:

Angle	°	′	″
BX	153	14	47
BA	59	21	14
ABX = BA − BX	93	53	33

The angles for stations C and D are calculated in the same way.

As a field check, it is useful also to know the distances to the new point. For point B, the grid distance is given by $\sqrt{143.916^2 + 72.551^2} = 161.169$ metres. To convert this into a horizontal distance on the ellipsoid, we must divide by the local scale factor (0.999859, as calculated by Equation D.1; this gives

HORIZONTAL/~~VERTICAL~~ OBSVNS AT: Station B   GROUP: **A2** | PAGE

Date: _____ Instrument: _____ Observer: _____   JOB: **Setting**

Time: _____ Ht of inst: _____ Booker: _____   **out**

Weather: _____ Checker: _____

Temperature (°C): _____ Pressure (mm Hg): _____

Face / swing	Stations and points	Observed Angle ° ' "	Mean F1/F2 ° ' "	Raw/Corrected HD/SD (delete as necessary)
F1 SR	Station A (R.O.)	00 00 00		
	Station X	93 53 33		161.169
	R.O. (check)			
			360° - VA	Target height
F2 SL				
	R.O. (check)			
	Station X	93 53 33		
	Station A (R.O.)	00 00 00		

Figure F.3 Booking data for setting-out (digital instrument).

161.192 metres. This can also be accepted as the horizontal distance without further calculation, as the ellipsoidal heights of all the points are small. Again, the distances from points C and D are calculated in the same way.

A sample booking sheet, prepared for setting out the angle from point B to point X using A as a reference object, is shown in Figure F.3. The blank areas in the form will be filled out in the field, as described in Chapter 4, Section 4.4.3.

F.2 CALCULATION USING LSQ

The above calculation can be done more quickly (and with fewer chances of mistakes) using LSQ. If the 'Data: Update' command is applied after the adjustment shown in Figure E.8, the 'guessed' co-ordinates for stations C and D in the input file are replaced with the calculated ones. The input file can then be manually edited in LSQ's built-in editor, as follows:

1. Set the co-ordinates of C and D to be 'Fixed'.
2. Add station X and its co-ordinates, also 'Fixed'.

3. Remove all the original observations, and replace them with 'dummy' observations, representing the readings whose values we need to know, i.e. the horizontal angles *ABX*, *ACX* and *FDX* plus the horizontal distances *BX*, *CX* and *DX*.
4. Make an appropriate change to the title.

After these changes, the input file will look as shown in Figure F.4. Note that any value can be used as an 'observed' horizontal angle (zero has been used in this case) but that observed distances cannot be zero, so a nominal '1' has been used here. Likewise, the ESDs are dummy values, but must be positive to be legal.

Loading this file into LSQ and simply viewing the results (adjusting the data has no effect, so is unnecessary) gives the output shown in Figure F.5. The relevant information is the set of 'CALCULATED' values for each of the dummy observations, which shows what angles and distances LSQ would have expected from the given input data.

As can be seen, the angle *ABX* and the distance *BX* are the same as in the manual calculation above; and all the other angles and distances have been calculated as well.

```
Data for setting out X
Projection BNG

A        545490.8400 F      257766.5900 F      10.0000 F
B        544777.4490 F      257343.9160 F      10.0000 F
E        545070.4280 F      257348.3710 F      10.0000 F
F        544988.9270 F      257720.0980 F      10.0000 F
C        544896.9245 F      257367.6011 F      10.0000 F
D        544979.9951 F      257344.2669 F      10.0000 F
X 544850 F   257200 F 10 F

HA A B X 0 0 0 1
HA A C X 0 0 0 1
HA F D X 0 0 0 1

HD B X 1 1
HD C X 1 1
HD D X 1 1
```

Figure F.4 LSQ input data after editing for setting-out.

```
Data for setting out X

* Program: LSQ V-8.80   Date: 15/05/2013    Time: 11:40:54   Cycles: 0
* Projection: BNG (British National Grid)  Ellipsoid: Airy 1830
Projection BNG  Airy  0

* Station co-ordinates (Adjustable co-ordinates updated)    Error ellipses
*        EASTING          NORTHING          HEIGHT       MAJOR  MINOR  AZ  HEIGHT
A      545490.8400  F    257766.5900  F    10.0000  F
B      544777.4490  F    257343.9160  F    10.0000  F
E      545070.4280  F    257348.3710  F    10.0000  F
F      544988.9270  F    257720.0980  F    10.0000  F
C      544896.9245  F    257367.6011  F    10.0000  F
D      544979.9951  F    257344.2669  F    10.0000  F
X      544850.0000  F    257200.0000  F    10.0000  F

* Analysis of readings:
* REF INST OBS       OBSERVED   ESD    HT-HI    CALCULATED      DIFF   WDIFF  ID
HA  A   B    X      0  0  0.00  1.00          |  93 53 32.90  ********  ******  1
HA  A   C    X      0  0  0.00  1.00          | 139 32  2.23  ********  ******  2
HA  F   D    X      0  0  0.00  1.00          | 220 39 34.95  ********  ******  3
HD      B    X         1.0000   1.000         |    161.1921  ********  160.19  4
HD      C    X         1.0000   1.000         |    174.0709  ********  173.07  5
HD      D    X         1.0000   1.000         |    194.2226  ********  193.22  6

ESD Scale Factor = 555971.279 after 0 cycles

Biggest errors:
WDIFF 502322.23  501625.05  338012.90    193.22    173.07    160.19
Obs ID      2          3          1          6         5         4
Status
Rejection limit: (None)
```

Figure F.5 LSQ results for setting out.

Appendix G: Booking Sheets

These sheets were developed for use at Cambridge University, and may be freely copied.

HORIZONTAL/VERTICAL OBSVNS AT: _____	GROUP:_____	PAGE

Date: _____ Instrument: _____ Observer: _____ JOB: _____
Time: _____ Ht of inst: _____ Booker: _____
Weather: _____ Checker: _____
Temperature (°C): _____ Pressure (mm Hg): _____

Face/ swing	Stations and points	Observed Angle ° ' "	Mean F1/F2 ° ' "	Raw/Corrected HD/SD (delete as necessary)
			360° - VA	Target Height

Figure G.1 Booking sheet for total stations.

HORIZONTAL/VERTICAL OBSVNS AT: _____				GROUP:___	PAGE
Date: _____ Instrument: _____		Observer: _____		JOB:	
Time: _____ Ht of inst: _____		Booker: _____			
Weather: _____		Checker: _____		_____	
Circle / swing	Stations and points	Observed Angle ° ' "	Reduced Angle ° ' "	Mean CL/CR ° ' "	
				HD / SD / Target Height	

Figure G.2 Booking sheet for theodolites.

LEVELLING From: _____		To: _____			GROUP:___	PAGE
Instrument: _____		Observer: _____			JOB: _____	
Date: _____		Booker: _____				
Weather: _____		Checker: _____				
Staff Position	Distance to Foresight	Foresight	Reduced Level	Distance to Backsight	Backsight	Line of Collimation
	→	→				

Figure G.3 Booking sheet for levels.

PRECISE DIGITAL LEVELLING OBSERVATIONS

From: _____ To: _____ Date: _____

Instrument: _____ Staves: _____ Weather: _____

Observer: _____ Booker: _____ Staffholder: _____

St 1 → St 2	Backsight	Foresight	Distance (m)		Diff in elev (m)	
Ht 1 Temp	(m)	(m)	Back	Fore	Rise	Fall
Totals				Totals		
Difference of Σ's			Difference of Σ's			
Level of		m		x ½		
Level of		m	Difference of means			

Notes: Minimise delay between backsight and foresight observations. Keep running totals of distances nearly equal. With 2 staves, each bay of the line must have an even number of instrument positions. Mark one of the staves • and always read it first.

Order of reading and booking:

1	3	2	4	(1)-(3)	
6	5			(6)-(5)	
Sum	Sum	Cum Tot	Cum Tot	Mean	
9	7	10	8		(7)-(9)
11	12				(12)-(11)
Sum	Sum	Cum Tot	Cum Tot		Mean

Figure G.4 Booking sheet for precise digital levelling.

GNSS OBSERVATIONS

Date: _____   Type of observation: _____   Card name: _____

Observer: _____   Antenna type/Serial no: _____

Station name	Antenna Mount A to E see↓	Start		Interval (secs)	Finish		Comment
		Time	Height see↓		Time	Height	

Mounting systems and offsets for AT502 antenna:

Type	Name	Record height (above, in metres) to: ** Do not add in any offset! **	Offset (metres) (Office use only)
A	AT502 Beacon	top of silver mounting post	0.010
B	AT502 Carrier	lower shoulder of carrier	0.110
C	AT502 Pillar	threaded silver base of antenna	0.000
D	AT502 Pole	default height is 2.000m (includes grey adapter)	0.000
E	AT502 Tripod	white line on height hook	0.360

Figure G.5 Booking sheet for GNSS observations.

RECIPROCAL VERTICAL ANGLE OBSERVATIONS

At (station): _____   To (station): _____

On (Main/Aux): _____   To (Main/Aux): _____

Instrument: _____   Observer: _____

Date: _____   Booker: _____

Weather: _____   Checker: _____

No.	Time	Zenith Angles		360-②	③-①	Mean of	④-④'
		CL/F1 ①	CR/F2 ②	③	④	① & ③	
		° ′ ″	° ′ ″	° ′ ″	″	° ′ ″	

Figure G.6 Booking sheet for reciprocal vertical angles (main observations).

HEIGHT INFORMATION

If the centre of the target is higher than the trunnion axis of the instrument, complete **Section 1**, otherwise complete **Section 2**. To reduce error the horizontal line of sight shown in the diagrams should be set using face 1 (90°) before the observations are made and face 2 (270°) afterwards.

1 TARGET HIGHER

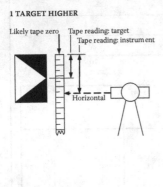

	Before m	After m	Mean m
A. Tape reading: instrument cross-hair			
B. Tape reading: centre of target			
C. Height of target centre above trunnion axis of instrument. Calculate: |A – B|			
D. Instrument height above main station Measure if instrument is directly over main station, otherwise calculate: **E – C**			
E. Target height above main station Measure if target is directly over main station, otherwise calculate: **D + C**			

2 TARGET LOWER

	Before m	After m	Mean m
A. Tape reading: instrument cross-hair			
B. Tape reading: centre of target			
C. Height of trunnion axis of instrument above target centre. Calculate: |A – B|			
D. Instrument height above main station Measure if instrument is directly over main station, otherwise calculate: **E + C**			
E. Target height above main station Measure if target is directly over main station, otherwise calculate: **D – C**			

STATION LAYOUT (OFFSET STATIONS)

Complete the plan view below by ticking the boxes to show the positions of the main stations, auxiliary stations, instruments and targets.

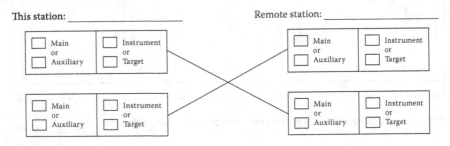

This station: _____   Remote station: _____

☐ Main or ☐ Auxiliary   ☐ Instrument or ☐ Target

☐ Main or ☐ Auxiliary   ☐ Instrument or ☐ Target

☐ Main or ☐ Auxiliary   ☐ Instrument or ☐ Target

☐ Main or ☐ Auxiliary   ☐ Instrument or ☐ Target

Figure G.7 Booking sheet for reciprocal vertical angles (station heights and layout).

Appendix H: Calculation Sheets

These sheets were developed for use at Cambridge University and may be freely copied.

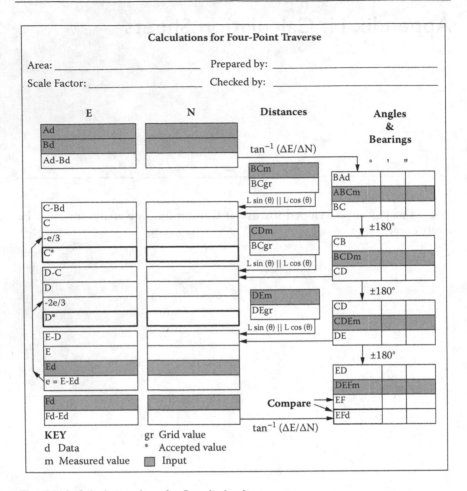

Figure H.I Calculation sheet for Bowditch adjustment.

Summary Sheet for Slope Distance Measurements

Observing Station: _____

Ellipsoidal height (H_1, metres): _____ Instrument: _____

Prepared by: _____ Checked by: _____

Name of observed (target) station	2					
Date of observation						
Time of observation						
Uncorrected observed distance	D_0	m				
Zero Correction (see table below)	C_0	m				
$D_0 + C_0$	D_1	m				
Mean temperature along light path	T	°C				
Mean pressure along light path	P	mm Hg				
Atmospheric correction (Formula A)	C_1	ppm				
$D_1 \times (1 + (C_1 \times 10^{-6}))$	D_2	m				
Refraction const. (visible light, if measured)	κ					
Propagation correction (Formula B)	C_2	m				
Relative altitude of station 2 (nearest cm)	$H_2 - H_1$	m				
Instrument ht above station 1 (nearest mm)	h_1	m				
Target height above station 2 (nearest mm)	h_2	m				
Height correction (Formula C)	C_3	m				
Accepted Slope Distance ($D_2 + C_2 + C_3$)		m				

Instrument	Wave-length	C_0	Formula A (ppm)	Accuracy
AGA 14A	infra-red	0	$275 - (106P / (273 + T))$	10mm + 3ppm
AGA 120	infra-red	0	$275 - (106P / (273 + T))$	5mm + 7ppm
TC-403L	infra-red	0	$282 - (105.5P / (273 + T))$	3mm + 3ppm
TC-405L	infra-red	0	$282 - (105.5P / (273 + T))$	2mm + 2ppm

Formula **B**: $C_2 = \dfrac{D_2{}^3}{24R_E{}^2} \times (2\kappa - \kappa^2)$ with κ as above or $\dfrac{1}{7}$ for infra-red, $\dfrac{1}{4}$ for microwaves

and R_E = Earth's radius = 6.371×10^6 m

Formula **C**: $C_3 = -\left(\dfrac{(H_2 - H_1) \times (h_2 - h_1)}{D_2} + \dfrac{D_2 \times (h_2 - h_1)}{2R_E} \right)$

Notes:

1. "nearest cm" or similar indicates the accuracy typically required for 1ppm accuracy in this calculation, not how a more accurate figure should be rounded.

2. $2R_E$ in formula C should be replaced by $2(R_E + H_1)$ if the ellipsoidal height of the observing station is greater than ~5000 metres

Figure H.2 Calculation sheet for slope distance measurements.

Calculations for Reciprocal Vertical Angles

Prepared by:_____   Checked by:_____

		Observing from station 1	Observing from station 2
Name of observing station			
Name of observed station			
Observing from (main/aux)			
Observing to (main/aux)			
Height of instrument above **observing** station ①	m		
Height of target above **observed** station ②	m		
Height difference [②−①] ③	m		
Approx. observed zenith angle (nearest minute) ④			
Slope distance (nearest cm) (Use D_2 from distance sheet) **(sd)**	m		
Correction to be added *	"		

* Correction (in seconds) is: $180 \times 3600 \times ③ \times \sin(④)/(sd \times \pi)$

Zenith Angles				Excess ⑤+⑥−180°	Difference ⑥−⑤
At station 1		At station 2			
Observed ° ' "	Corrected ⑤ ° ' "	Observed ° ' "	Corrected ⑥ ° ' "	° ' "	° ' "

Mean Excess in seconds **(xs)**:_____

Calculated refraction constant (κ)

$$= 1 - \frac{R_E \times xs \times \pi}{sd \cos(SA) \times 180 \times 3600} : \underline{\hspace{2cm}}$$

where R_E = Earth's radius $\approx 6.371 \times 10^6$ m (Compare with standard refraction constant for visible light, $\kappa = 1/7 \approx 0.1429$)

Mean Difference **(MD)**

Slope Angle $\left[\dfrac{MD}{2}\right]$ **(SA)**

Approx. height of station 2 above station 1 (to 1 cm) [$sd \times \sin(SA)$] **($H_2 - H_1$)**

Exact Δh (using fully corrected slope distance)

Note: "nearest minute" (or similar) is an indication of the minimum accuracy required, not an indication of how a more accurate value should be rounded.

Figure H.3 Calculation sheet for reciprocal vertical angles.

References and Bibliography

GENERAL

Allan, A.L. (1997) *Practical Surveying and Computations*, 2nd ed., Oxford Laxton's. ISBN 0750636556. Good easy-to-follow coverage of the mathematical formulae and calculations used in surveying.

Bannister, A., Raymond S. and Baker, R. (1998) *Surveying*, 7th ed., Harlow: Addison Wesley Longman Ltd. Covers some traditional surveying methods well, including chaining and taping.

Bomford, G. (1980) *Geodesy*, 4th ed., Oxford: Clarendon Press. ISBN 019851946x. Now out of print but available in most good technical libraries, and still a valuable reference text for many aspects of surveying.

Schofield, W. and Breach, M. (2007) *Engineering Surveying*, 6th ed., Oxford: Butterworth-Heinemann. ISBN 0750669497. Good coverage of setting-out techniques for building sites and underground surveying, including the use of gyrotheodolites.

Uren, J. and Price, W.F. (2006) *Surveying for Engineers*, 4th ed., Basingstoke: Palgrave MacMillan. ISBN 1403920540. Good description of setting out road lines, and earthworks calculations.

DISTANCE MEASUREMENT

Burnside, C.D. (1991) *Electromagnetic Distance Measurement*, 3rd ed., Oxford: BSP Professional. ISBN 0632031220. Concentrates mainly on how the earlier generations of instruments work

Rüeger, J.M. (1996) *Electronic Distance Measurement*, 4th Ed., Springer-Verlag. ISBN 3540611592. Very thorough treatment of corrections to distance measurements.

MAP PROJECTIONS

Bugayevskiy, L.M., and Snyder, J.P. (1995) *Map Projections: A Reference Manual*, London: Taylor & Francis. ISBN 0748403043. A comprehensive and mathematical description of a wide range of projections.

Iliffe, J.C. (2000) *Datums and Map Projections*, London: CRC Press. ISBN 0849308844. Shows clearly how to choose and use a projection in conjunction with GPS data.

Jackson, J.E. (1987) *Sphere, Spheroid and Projections for Surveyors*, Oxford: BSP Professional. ISBN 0632018674.

Maling, D.H. (1993) *Co-ordinate Systems and Map Projections*, 2nd ed. (corrected), Oxford: Pergamon Press. ISBN 0080372333.

Ordnance Survey (1950) *Constants, Formulae and Methods used in the Transverse Mercator Projection*, London: HMSO. ISBN 0-1170-0495-2. Still the authoritative publication from the Ordnance Survey on this subject, and available as a 'print on demand' title.

PHOTOGRAMMETRY

Atkinson, K.B. (2001) (Editor) *Close Range Photogrammetry and Machine Vision*, Caithness: Whittles Publishing. ISBN 1870325737. A useful selection of papers covering the theory and application of photogrammetry in industry, architecture and medicine.

Egels, Y. and Kasser, M. (2001) *Digital Photogrammetry*, London: Taylor & Francis. ISBN 0748409440.

Mikhail, E.M., Bethel, J.S. and McGlone, J.C. (2001) *Introduction to Modern Photogrammetry*, New York: Wiley. ISBN 0471309249.

Wolf, P.R. (2000) *Elements of Photogrammetry: with Applications in GIS*, London: McGraw Hill. ISBN 0072924543.

SATELLITE METHODS

Leick, A. (2004) *GPS Satellite Surveying*, 3rd ed., New York: Wiley. ISBN 0471059307.

Van Sickle, J. (2001) *GPS for Land Surveyors*, 2nd ed., London: Taylor & Francis. ISBN 1575040751.

STATISTICS AND ADJUSTMENT

Wolf, P.R., and Ghilani, C.D. (1997) *Adjustment Computations: Statistics and Least Squares in Surveying and GIS*, 3rd ed., New York: Wiley-Interscience. ISBN 0471168335.

GEODESICS

Karney, C.F.F. (2012) Algorithms for Geodesics, *Journal of Geodesy* Vol 87 Issue 1, pp 43–55, available as an open access article from (April 2013) http://link.springer.com/article/10.1007%2Fs00190-012-0578-z. Gives a link to a library containing an implementation of an algorithm for calculating geodesics which can be freely downloaded.

Glossary

This section contains brief descriptions of terms which may be unfamiliar to the reader. See the index for a fuller discussion of each term in the main text.

Alidade bubble The bubble (usually a split bubble) used to set the vertical circle, usually so that the zero degree marker is pointing directly upwards.

Azimuth angle The angle measured clockwise at a point on the earth's surface from the direction of true north (i.e. along a meridian through the point) to another horizontal line passing through the point.

Backlash The looseness or 'play' in a mechanism which means that not all parts of the mechanism are always in the same place when one part of it is moved to a particular position. This can, for instance, affect the angle read from some types of theodolites for a given sighting, depending upon whether the final adjustments to the telescope and vernier were made in a clockwise or anticlockwise direction.

Backsight The sighting from a level to a staff positioned on a point whose height is known. The level's line of collimation can then be calculated. See Foresight.

Bay A sequence of levelling backsights and foresights which either closes on itself, or which runs from one point of known height to another. The 'closure' of the bay is an indication of the accuracy of the readings within it.

Bearing See Azimuth angle.

Bubble error The maladjustment of a spirit bubble such that a piece of equipment is not exactly horizontal (or vertical) when the bubble indicates that it is.

Change point The point occupied by a levelling staff when the instrument is moved, to level the next part of the bay.

Circle One of the two protractors in a total station or theodolite, on which horizontal or vertical angles are measured.

Clinometer A simple optical device incorporating a pendulum (or spirit level) and protractor, for estimating vertical angles.

Closure   A check on the consistency of a set of observations, e.g.,: (1) Re-observing a reference object after taking a set of horizontal angles. A reading which differs from the original reading by more than the accuracy of the instrument makes the other readings suspect. (2) Seeing whether the height of an object given by a set of levelling measurements corresponds with its known height.

Collimation, line of   The line of sight between the centre of the crosshairs in the eyepiece of a telescope and the distant object they appear to intersect, once parallax has been removed. In levelling, the height of that (horizontal) line.

Cup bubble   A circular 'spirit level' bubble, used for setting a tribrach or instrument approximately horizontal.

Detail pole   An extendable pole about 1–2 metres in length, equipped with a reflector, used for collecting detail for mapping. Colloquially known as a 'pogo'.

DGNSS   See Differential GNSS, below.

Differential GNSS   The simultaneous reception of satellite signals by two receivers, one of which is a known position. The position of the other receiver can then be calculated to high accuracy.

EDM   Electromagnetic distance measurement. Measurement of distance by counting the number of cycles (between a transmitter/receiver and reflector) of an electromagnetic wave whose wavelength is known; or by measuring the time taken for a laser pulse to be reflected back from a distant object.

Ephemeris   The orbital parameters of a satellite, which are used to determine its position in space at a given point in time.

EGNOS   The European Geostationary Navigation Overlay System, which enhances accuracy of the GPS system over Europe.

Epoch   An instant in time, e.g. 00:00 hours on January 1, 1989.

Error ellipse   An indication of the accuracy to which a point's position is known, following least-squares adjustment.

Face   An alternative name for the vertical circle on a total station or theodolite.

Foresight   The sighting from a level (whose line of collimation is known) to a staff positioned on a point whose height is required. See Backsight.

GDOP   Geometric Dilution Of Position. The ration between the accuracy with which a GNSS receiver can determine its location and the accuracy with which it can measure its distance from the satellites it is observing. Its value is affected by the positions of the observed satellites in the sky, and their estimated clock errors.

Geodesic   The shortest path over the surface of an ellipsoid between two points on that surface.

Geoid   The irregularly-shaped surface defined by the locus of all points around the earth with the same gravitational potential as some given datum; in the British Isles, the datum is a marker near the mean sea level at Newlyn, in Cornwall.

Geodetic   The co-ordinate system which defines the position of a point by quoting its latitude, longitude and distance above the surface of an ellipsoid.

Geomatics   A term sometimes used to describe the type of surveying covered by this book, emphasising the elements of geodesy and informatics used within the discipline.

GIS   Geographic Information System. Any system for organising geographic data to inform decision-making. This includes, but is not limited to, the spatial data which is collected to help plan engineering works.

GLONASS   Globalnaya NAvigatsionnaya Sputnikovaya Sistema (or Global Navigational Satellite System): the network of navigational satellites managed by the Russian Aerospace Defence Forces, which works in a similar way to GPS (see below).

GNSS   Global Navigational Satellite System: A collective term for all the available navigational satellite systems, such as GPS (see below) and GLONASS (see above).

GPS   Global Positioning System: the network of satellites managed by the USA, which enables the position of a station to be determined by measuring its distance from four or more of them.

HDOP   Horizontal Dilution Of Positions. The horizontal elements of PDOP (see below), used to estimate the accuracy with which a GNSS receiver can measure its position in the horizontal plane.

Homogeneous   In the context of surveying, this means 'only containing the mathematically predictable distortions inherent in the system of projection'.

Horizontal circle   See Circle.

Level   A telescope designed to have a horizontal line of collimation, used for comparing the heights of two stations.

Meridian   A line of constant longitude on the earth's surface; the line produced when the earth is cut by a planar surface which contains the polar axis.

Navigational GNSS   The use of GNSS data to calculate the position of a single receiver, without reference to any other receivers.

Orthometric height   A height measured from the local geoid.

Parallax   The spatial separation between, for instance, the image of a distant object and the crosshairs in an instrument, which causes the two to appear to move relative to each other depending on the position of the observer's eye.

Parallel   A line of constant latitude on the earth's surface; the line produced when a planar cut is made through the earth normal to the polar axis.

PDOP   Positional Dilution Of Position. Similar to GDOP (see above) but based soley on the positions of the satellites. Typically quoted in place of

GDOP when insufficient satellites are visible for their clock errors to be estimated.

Photogrammetry A technique for determining the relative three-dimensional positions of points by taking photographs of the points from two different places. Two overlapping photographs taken from an aeroplane or satellite can be used to create topographic maps by this technique.

Plate An alternative term for the horizontal circle on a total station or theodolite.

Plate bubble A levelling bubble, usually tubular, built into a total station or theodolite to set its plate (i.e. horizontal circle) level.

Prime vertical The plane passing through a point on the ellipsoid which contains the surface normal and the east–west vector passing through the point. It can also be defined as the plane which is perpendicular to the tangent plane and the plane of the meridian at the point.

Ranging rod A striped pole, pointed at one end, used for checking lines of sight. The stripes are usually a decimetre wide, enabling the rod to be used as a crude 'ruler'.

Real-time kinematic A method of differential GNSS survey in which the two receivers are in radio contact with each other, and can thus calculate the exact difference in their positions in real time.

Redundancy The principle of taking more measurements than are strictly required to fix the positions of unknown points, so that any errors in the measurement data become apparent. This is analogous to a 'redundant' structure, which has more bracing than necessary to prevent it falling down.

Reciprocal vertical A method of observing vertical angles simultaneously from both ends of a ray, to cancel out atmospheric effects.

Reference object A station observed at the start and end of a set of horizontal angle observations, whose reading on the horizontal circle is used as a datum for the other readings. The co-ordinates of the reference object need not necessarily be known at the time of observation—it is more important that the observation should be one which it is easy to take consistently.

Resectioning Establishing the position of a station by mounting an instrument over it, and measuring angles and distances to other known stations.

RINEX Receiver-INdependent EXchange. A format used to transfer GNSS data between receivers and post-processing software.

Round A complete set of horizontal angles, measured circle left and circle right, of one or more stations, with respect to a reference object.

RTK See Real-time kinematic.

RV See Reciprocal vertical.

Scale factor The factor by which the actual scale of a map at a given locality differs from its nominal scale. A local scale factor of, say, 1.25 on a 1:50,000 map would give a local scale of 1:40,000.

Slope angle   The angle between a sloping line and a horizontal plane.

Split bubble   A bubble with an optical arrangement which shows its opposite ends side by side. The bubble is centred by lining up the images of the two ends. This is more precise and repeatable than using engraved marks on the glass, but it does not eliminate bubble error.

Stadia hairs   Two (usually horizontal) hairs parallel to the main crosshairs in a telescope, subtending a fixed angle of observation, and used in tachymetry.

Staff   A graduated rod, used for levelling and tachymetry.

Station   A fixed point on the ground, whose position is known or required.

Staves   The plural of staff.

Swing   The term for rotating a total station or theodolite about its vertical axis. 'Swing left' means rotate anticlockwise, as seen from above.

Tachymetry   The measurement of distance by observing the length along a distant staff which subtends a known angle at the instrument, usually by using the stadia hairs in the telescope.

Tangent screw   An adjustment screw tangential to the horizontal or vertical axis of an instrument, which enables precise aiming of the telescope.

Target   A visual reference mounted over a station, designed for precise and repeatable observations of angles to that station.

Temporary benchmark   A point established at (usually) the extreme end of a levelling bay, which will be used as the starting point for another bay.

Theodolite   A telescope equipped with protractors in the horizontal and vertical planes, capable of precise measurement of azimuth and elevation angles.

Total station   An integrated theodolite and EDM device which can measure angles and distances, and often has the capability to record readings electronically.

Transit   The act of turning the telescope on a total station or theodolite through the vertical, e.g. when changing from face 1 to face 2.

Traverse   The establishment of successive station co-ordinates by finding bearings and distances from a previous station. A traverse is usually 'closed' by using a known station as the final one.

Tribrach   An adjustable platform which fixes to a tripod, and provides a location for an instrument or target which is level, and directly above a station. Tribrachs have footscrews for levelling, and may incorporate a plate bubble and an optical or laser plummet.

TRF   Terrestrial Reference Frame. A set of fixed points on the earth's surface, with published co-ordinates, which effectively define the TRS (see below) with respect to the earth. Inevitably, there will be small inconsistencies in a TRF due to the impossibility of measuring the relative positions of the points exactly, plus the fact that the points may move relative to one another after the measurements have been made.

TRS   Terrestrial Reference System. A set of Cartesian axes, with an associated ellipsoid of defined size and shape, whose position with respect to the earth is defined in some way. The system may also include a specified method for updating the co-ordinates of fixed stations, when more accurate measurements of their relative positions become available.

Trunnion axis   The horizontal axis in a total station or theodolite; the bearings which support the telescope.

VDOP   Vertical Dilution Of Position. The vertical element of PDOP (see above), used to estimate the accuracy with which a GNSS receiver can measure its ellipsoidal height.

Vernier   A mechanism for reading a scale to greater precision. An optical vernier bends a light path until two markers line up, and then indicates the amount by which the path was bent.

Vertical circle   See Circle.

Zenith angle   The angle between a line of sight and the vertical at the point of observation.

Index

Printed in the United States
by Baker & Taylor Publisher Services